T0133421

Plantations, Privatization, Poverty and Power

Changing Ownership and Management of State Forests

Edited by
Mike Garforth
and
James Mayers

London • Sterling, VA

THE EARTHSCAN FORESTRY LIBRARY
Plantations, Privatization, Poverty and Power:
Changing Ownership and Management of State Forests
Mike Garforth and James Mayers (eds)

The Sustainable Forestry Handbook 2nd edition
Sophie Higman, James Mayers, Stephen Bass, Neil Judd and Ruth Nussbaum

The Forest Certification Handbook 2nd edition
Ruth Nussbaum and Markku Simula

First published by Earthscan in the UK and USA in 2005

ISBN: 1-84407-152-9 paperback
1-84407-151-0 hardback

Typesetting by JS Typesetting Ltd, Porthcawl, Mid Glamorgan
Printed and bound in the UK by Cromwell Press, Trowbridge
Cover design by Anthony Waters

For a full list of publications please contact:

Earthscan
8–12 Camden High Street
London, NW1 0JH, UK
Tel: +44 (0)20 7387 8558
Fax: +44 (0)20 7387 8998
Email: earthinfo@earthscan.co.uk
Web: **www.earthscan.co.uk**

22883 Quicksilver Drive, Sterling, VA 20166-2012, USA

Earthscan is an imprint of James and James (Science Publishers) Ltd and publishes
in association with WWF-UK and the International Institute for Environment and
Development

A catalogue record for this book is available from the British Library

Library of Congress Cataloging-in-Publication Data

Plantations, privatization, poverty and power : changing ownership and
management of state forests / edited by Michael Garforth and James Mayers.
    p. cm.
  Includes bibliographical references and index.
  ISBN 1-84407-151-0 (hardback) – ISBN 1-84407-152-9 (pbk.)
  1. Forest reserves–Management. 2. Tree farms–Management. I. Garforth,
Michael. II. Mayers, James.
    SD561.P58 2005
    333.75'11–dc22

                                    2004024816

Printed on elemental chlorine-free paper

Contents

List of Figures, Tables and Boxes

Figures

Tables

Boxes

List of Acronyms and Abbreviations

ABARE	Australian Bureau of Agricultural and Resource Economics
ABS	Australian Bureau of Statistics
ACF	Australian Conservation Foundation
ACT	Australian Capital Territory
AFFA	Agriculture Fisheries and Forestry, Australia
ANC	African National Congress
ANCOM	Andean Common Market
APS	Afforestation Permit System, South Africa
ASIMAD	Asociación de Industriales de la Madera, Chile
ATCP Chile	Asociación Técnica de la Celulosa y el Papel de Chile, Chile
ATO	Australian Tax Office
CAP	Common Agricultural Policy
CBD	Convention on Biological Diversity
CCTV	China Central Television
CFL	Crown Forestry Licence, New Zealand
CHH	Carter Holt Harvey, New Zealand
CIPMA	Centro de Investigación y Planificación del Medio Ambiente, Chile
CoA	Commonwealth of Australia
CODEFF	Flora and Fauna Defence Committee, Chile
CONAF	Corporación Nacional Forestal, Chile
CONAMA	Comisión Nacional de Medio Ambiente, Chile
COREF	Corporación de Reforestación, Chile
CORFO	Corporación de Fomento de la Producción, Chile
CORMA	Corporación Chilena de la Madera, Chile
COSATU	Congress of South African Trade Unions
CRC–SPF	Cooperative Research for Sustainable Production Forestry, Australia
CRS	Contract Responsibility System, China
cum	cubic metres
DEAT	Department of Environmental Affairs and Tourism, South Africa
DFID	Department for International Development, UK
DLWC	Department of Land and Water Conservation, Australia, New South Wales

DoC	Department of Conservation, New Zealand
DPE	Department of Public Enterprises, South Africa
DWAF	Department of Water Affairs and Forestry, South Africa
ECA	European Community Accession
ECFP	East Coast Forestry Project, New Zealand
EMV	expected market value
ENGO	environmental non-governmental organization
ESOP	Employee Share Ownership Plan, South Africa
EU	European Union
FAFPIC	Forestry and Forest Products Industry Council, Australia
FAO	Food and Agriculture Organization of the United Nations
FC	Forestry Commission, UK
FD	state forest department, India
FDA	forest development agency, India
FDC	forest development corporation, India
FE	Forest Enterprise, UK
FFPC	Forestry and Forest Products Council, Australia
FPB	Forest Practices Board, Australia
FPC	forest protection committee
FPCom	Forest Products Commission, Australia
FSC	Forest Stewardship Council
FSI	Forest Survey of India
FUG	Forest User Group, Nepal
FWPRDC	Forest and Wood Products Research and Development Corporation, Australia
GATT	General Agreement on Tariffs and Trade
GDP	gross domestic product
GM	genetically modified
GNP	gross national product
GoI	government of India
GTE	government trading enterprise, Australia
ha	hectares
HRS	Household Responsibility System, China
HVP	Hancock Victorian Plantations Pty Ltd, Australia
ICEFI	Independent Initiative in Forest Certification, Chile
ICFRE	Indian Council of Forest Research Education and Training, India
IIED	International Institute for Environment and Development
INDAP	Instituto de Desarrollo Agropecuario, Chile
INFOR	Instituto Forestal, Chile
INTEC	Instituto de Investigaciones Tecnologicas, Chile
ISO	International Organization for Standardization
JFM	Joint Forest Management, India
km	kilometres
LIRO	Logging Industry Research Organization, New Zealand
LTFT	Lake Taupo Forest Trust, New Zealand

m	metres
MAF	Ministry of Agriculture and Forests, New Zealand
MDBC	Murray–Darling Basin Commission
mm	millimetres
MoEF	Ministry of Environment and Forests, India
MoF	Ministry of Forestry, New Zealand
MW	megawatt
NAFTA	North American Free Trade Agreement
NCA	National Commission on Agriculture, India
NCP	National Competition Policy, Australia
NEF	National Empowerment Fund, South Africa
NFAP	National Forestry Action Programme, India
NFI	National Forest Inventory, Australia
NFPP	Natural Forest Protection Programme, China
NGO	non-governmental organization
NNFFR	North/North-west Farm Forest Region, China
NRRPC	Northern Rivers Regional Plantation Committee, Australia
NSSFR	North-east/South-west State Forest Region, China
NSW	New South Wales, Australia
NTFP	non-timber forest product
NWDB	National Wastelands Development Board, India
NZ$	New Zealand dollars
NZFC	New Zealand Forestry Corporation
NZFOA	New Zealand Forest Owners Association
NZFP	New Zealand Forest Products
NZFS	New Zealand Forest Service
NZIER	New Zealand Institute of Economic Research
PEFC	Programme for the Endorsement of Forest Certification
PPP	personal purchasing power
PRI	*Panchayati Raj* institution, India
R&D	research and development
RMA	Resource Management Act, New Zealand
RPC	regional plantation committee, Australia
RSA	Republic of South Africa
SA	South Australia
SAFCOL	South African Forestry Corporation
SAG	Servicio Agrícola y Ganadero, Chile
SCFR	Southern Collective Forest Region, China
SEPA	State Environment Protection Agency, China
SERCOTEC	Servicio de Cooperación Técnica, Chile
SFA	State Forest Administration, China
SFNSW	State Forests New South Wales, Australia
SNASPE	State Protected Wildlife Areas System, Chile
SOE	state-owned enterprise, China
SOFE	state-owned forest enterprise, China
SPV	special purpose vehicle, South Africa

TEPCO	Tokyo Electric Power Company
TVE	township and village enterprise, China
UK	United Kingdom
UN	United Nations
UNCED	United Nations Conference on Environment and Development
UNFCCC	United Nation's Framework Convention on Climate Change
UPPS	Unified Procurement Pricing System, China
US	United States
UTM	unidad tributaria mensual
VDC	village development committee, Nepal
ViPC	Victoria Plantation Corporation, Australia
VPC	Village Protection Committee, India
WA	Western Australia
WB	World Bank
WEC	World Energy Council
WRDP	Western Regional Development Programme, China
WTO	World Trade Organization
WWF	World Wide Fund for Nature

Executive Summary

Introduction: Plantations and change

Over the last century, the area of forest plantations globally increased from an insignificant area to 187 million hectares. More than 4 million hectares are being added annually. Depending on their design, plantations can provide an array of goods, services and downstream benefits as varied as natural forests, though they cannot substitute for all natural forests' values. Governments have played a major role as plantation developers and managers, and as promoters and subsidizers of private-sector plantation investment. The benefits actually sought by governments and other players have varied with countries' economic, social and environmental states and trends, influenced by competing domestic interests and comparative advantage in production and processing.

Increasingly, governments have been seeking to reduce their own direct involvement in plantation management. Tenure and use rights over the assets of state plantation enterprises are being transferred into private hands, and government agencies are increasingly outsourcing plantation operations and support services.

Motivations for this have varied from one country to another. Some are common – a drive for greater efficiency and profitability and reduced pressure on the public purse. Others are unique to specific circumstances, such as improving rural livelihoods and empowering disadvantaged groups. Different models have been tried; some have worked well in terms of delivering the desired outcomes, others less well. Many factors, in the design of the approach, the handling of the transfer process and the ongoing management of the outcomes of the process, determine success or failure.

In this book we retell the experience of seven countries: Australia, Chile, China, India, New Zealand, South Africa and the UK, and draw from their experiences and from wider experience to identify the opportunities that changing ownership and management offers to governments and the people whom they serve, as well as the challenges and solutions. We look at what has worked and what has failed to deliver the objectives of change, and what has created unexpected or unmanageable problems. We attempt to develop some general guidance on good practice in balancing the objectives of the different players to get a solution that works for plantations and people.

Changing drivers of plantation development

Governments' support for and direct involvement in plantation development has been responsible for creating the lion's share of the global total, acting by land purchase or lease, profit-sharing partnerships with private landowners, or through supportive policy, financial incentives and tax incentives. Governments' objectives are usually mixed, and the balance between objectives has changed as societies' needs and values have evolved. We can expect this balance to continue to change as societies evolve.

Government motivations have included the following:

- *Creating a national strategic reserve of timber* was important for several countries until the middle decades of the 20th century when the opening up of international trade and recognition of countries' comparative advantages in wood production and processing removed it from most policy agendas.
- *Promoting industrial development*: in addition to supporting local industries, timber was seen by many as a key export, providing valuable foreign exchange.
- *Creating employment and rural livelihoods* was an early driver in many countries and, although diluted in its impact by mechanization in some countries (e.g. the UK and Australia), remains key in others (e.g. India, South Africa and China).
- *Growing sources of energy*: plantations have been developed explicitly as a source of fuelwood in India and China, and as an alternative to fossil fuels in the UK.
- *Responding to demands for environmental services*, such as flood prevention, biodiversity conservation, landscape enhancement, recreation, protecting soil and water (including demands to limit the water used by plantations in, for example, South Africa and India) and carbon sequestration, have stimulated governments to increase regulation and to force plantation managers to reduce the commercial productivity of their assets and improve compliance with environmental management standards.

Private-sector motivations

Private-sector investment in plantations has been encouraged by government incentives and supportive tax policies in many countries; but these have not been a pre-condition for private-sector plantation development everywhere. For example, traditional private estates in the UK have established plantations for game, timber and private amenity, while mine owners in Chile began establishing plantations to supply pit props during the early 1900s. As large multinational players have entered the arena, investment in plantation development for profit from wood sales or to provide raw material for vertically integrated business has emerged without government support. The trend is most apparent in countries with a comparative advantage owing to high growth rates and low land and labour costs. In some countries, private-sector motivation is about 'staying in the game' as government or market pressure requires companies to demonstrate responsible stewardship by certifying their performance. Some companies are recognizing new opportunities in emerging markets for environmental services, such as carbon sequestration.

Changing government roles

Governments have been the pioneers of plantation development. Real or perceived lack of private-sector capacity and technical knowledge, a tradition of government responsibility for and power over forests, and centrally planned governance paradigms have been at the root of government actions to establish and manage plantations.

Forestry departments have acted as moderators of competing demands and continue to do so through regulation, penalties for damaging impacts, land zoning for forestry and environmental assessment of plans and operations. They have harnessed other arms of government to support private players with financial incentives and tax regimes. The mix and design of instruments has evolved, with increasing emphasis on payments for public goods and away from direct financial support for timber production.

Governments have developed a variety of institutional models around their investment in state plantations: autonomous agencies, horizontal functional splits between the plantation development, and other functions of forestry departments and integration of the plantation function, with other functions right down to the local level. Just as objectives have changed, so too has institutional design as the conflicts between state agencies' roles of developer and manager and of moderator have come to the fore.

Today, the trend is firmly away from direct involvement except where the private sector is manifestly unable to deliver – for example, where the high costs and low financial returns from investment weighted towards the production of public goods act to exclude private developers. Even here, some well-crafted financial incentives aim to deliver the goods through the private sector more efficiently – for example, through competitive tendering for plantation establishment projects.

Opportunities and concerns in changing ownership and management

New opportunities

Transfer of tenure and use rights over state plantation assets and privatizing plantation operations and support services offer several opportunities, usually presenting themselves in combination. Some are complementary, others may be contradictory and require trade-offs to be made:

- *Increased economic efficiency and improved aggregate welfare.* This may be achieved where markets process information on society's demands and costs of production more effectively than governments, and where private individuals are better at determining production and consumption and are less prone to rent-seeking. Privatization may result in more efficient tree-growing, and may also be a lever for raising efficiency in the processing sector – for example, by increasing effective competition for raw material supplies. However, markets, to date, have operated less effectively when it comes to public goods and externalities.
- *Controlled budget deficits.* Unprofitable plantation operations contribute to worsening budget deficits with direct implications for the resources available to other sectors, such as health and education. Changing ownership may bring gains to the public purse, although these may be offset by the costs of ensuring that the new owners comply with forestry regulations and deliver public benefits. Such costs may include additional forest authority staff and payments to the new owners for environmental services.
- *Increased entrepreneurial drive and investment in the forest.* By encouraging private entrepreneurs into forestry, the government opens the door for increased innovation and longer-term growth. Engaging communities as partners can help to combat forest degradation and improve forest condition by encouraging more responsible stewardship.
- *Improved forest governance.* Government systems for ensuring accountability are not always effective. Forestry officials may not necessarily manage plantation resources in the national or local interest. Also, where governments' role of developer and moderator are combined in one entity, the commercial interests of the state agency may be put before the interests of private plantation owners who are competing in the same markets. Privatization offers a way of divorcing and clarifying governments' roles. It may also offer a route for returning power to local people and promoting a wider and more representative pattern of ownership. Transfers of rights and ownership can change governance relationships for the better – for example, by improving communities' capacities to plan and manage businesses and to act inclusively.
- *Poverty reduction and improved rural livelihoods.* Designed in the right way, asset transfers can increase income to rural households and communities, and can benefit disadvantaged groups, including women. Benefits may come from higher levels of sustainable production stimulated by ownership of the asset and the products derived from it. However, it is a major challenge to distribute benefits fairly. Accurate assessment of the links between forests, forest enterprise and local livelihoods are important, and transaction rules may be needed specifically to target poor groups and prevent capture of benefits by elites.

Concerns

Perceptions abound that private companies are not accountable to public demands and have no incentive to provide important environmental and social goods and services. The private sector, it is argued, answers only to shareholders and its chief aim is profit. Such concerns present legitimate challenges to processes of transferring ownership and management:

- *Economic concerns.* These stem from the threat of resource 'mining' and the loss of valuable timber and non-timber forest assets as companies seek to recover the costs of their investment in as short a time period as possible, and from the repercussions of downstream processing, where governments have supported fledgling industry with cheap logs or guaranteed supply.
- *Social concerns.* New owners may restrict access – for example, to communities living in or near plantation areas – impacting upon traditional use rights and livelihoods or conflicting with public recreational and spiritual pursuits. Where government-run plantations have been criticized for inefficiency and low profitability, privatization is often associated with the reduction in spare capacity and labour redundancies.
- *Environmental concerns.* The threat of plantations being converted to another use can give rise to major concerns where they are valued for landscape, biodiversity and other environmental services. Even where private plantations are maintained, forest survey and management planning may not put as much weight on the value of 'non-productive' assets (such as biodiversity and ancient monuments) as would local people and non-governmental organizations (NGOs). Pressure to drive down costs may lead new owners to cut corners, resulting in increased pollution from spillages or unsafe disposal of waste engine and lubricating oil or pesticides.

Of course, reality is far less clear-cut than is often portrayed by either the biggest fans or critics of transfers of plantation ownership. There is no absolute case for private or public control of specific goods and services: the public or private nature of goods and services is not static, but depends upon the level of institutional sophistication, communications and technology; the dichotomy often presented between public and private is misleading. Lessons learned from experience certainly suggest that unwanted outcomes from privatization can be avoided through good design and adequate safeguards against negative impacts. Safeguards may be embedded in the governance system or incorporated within contracts for service provision and transfer agreements.

The main transfer options available

Three main types of transfer can be distinguished: outsourcing of services, transfer of use rights and transfer of ownership. Outsourcing is the least dramatic; ownership and overall management control are retained by the state, while particular use and management functions are devolved to private contractors. Transfer of use rights involves a greater devolution of power from the state plantation manager to non-state entities than contracting out. With outsourcing of harvesting or management activities, the private contractor continues to work for the state plantation manager. Where use rights are transferred, private harvesters work for themselves. The ultimate form of privatization involves a transfer of ownership rights over some or all of the assets that comprise the plantation. The ownership rights of the state often limit the scope of any transfer – the state may own the trees but not the land, or the land and trees but not the rights to take game.

Outsourcing services

Outsourcing of plantation operations, such as establishment, felling and extraction, and service functions, such as seedling production, vehicle and machinery supply and maintenance, is an effective and efficient way of reducing the costs and increasing the profitability of plantation enterprises. This model tends to be favoured where public benefits are felt to be too important to risk handing outright to private operators (e.g. where a plantation is providing an essential biological corridor between remnant patches of natural forest), but where the state's performance in managing the resource has been substandard.

When capacity, skills and competition are lacking, they can be developed quickly through cash payments to support start-up or development of contracting and training organizations, transfers of machinery, or guarantees of work to take new contracting business through the start-up period. Outsourcing of management planning activities may also yield gains; but the evidence is weaker. A further step could be to contract out the whole management system for plantations; but it is not clear that there would be net gains, in particular where the forest is to be managed for multiple benefits and the management contract becomes complex in an attempt to specify in advance how, for example, production is to be traded against biodiversity over a long period.

Transfer of use rights

Governments may favour this model of transfer where they lack the managerial capacity to operate a sophisticated outsourcing system. While they must still monitor compliance with licence conditions, the burden is likely to be lower than that for outsourcing. The extra degree of freedom for the beneficiary is at least partly curtailed by the imposition of harvesting conditions. The simplest licence will limit the volume of wood extracted, while more complex agreements will set out how and when private operators are permitted to extract their timber products. Otherwise, many of the prerequisites necessary for effective outsourcing will still apply. Private-sector capacity remains a paramount challenge, as does a minimum level of competition. Use-rights transfers also pose a number of unique challenges for governments; these relate to the method of allocating use rights and the design of use rights, in particular their duration and transferability.

Transfer of ownership

The key question to ask in relation to this model is 'ownership of what?' The common assumption is that we are referring to ownership of land and everything on that land. In practice, however, transfers are often much more circumscribed. Where tree ownership is privatized, rights to the land, underlying minerals, environmental services and non-timber forest products may not be transferred, although local laws may be ambiguous on this point. In practice, the transfer of exclusive rights to all of the assets vested in state agencies is rare.

Key reasons governments may choose to transfer ownership rights include:

- *Efficiency.* There is a belief that partial rights transfers (e.g. of use or management rights) are less effective in stimulating entrepreneurial drive and improved efficiency amongst private actors.
- *Budgetary requirements.* The government's plantation authority may be severely short of cash and unable to continue to perform its responsibilities. Privatization both transfers its liabilities and cashes in on its plantation assets.
- *Limited environmental and social concerns.* Full ownership privatization is likely to be favoured in plantation areas where the public benefits attached to the plantations, and thus public resistance to their transfer, are relatively small.

Key challenges

Engaging local groups in ownership and management

For governments concerned with the social impacts of privatization in rural areas, a key challenge is ensuring local participation in both the process and the final outcome. Small local landholders face a number of hurdles in trying to take part. Common constraints revolve around their lack of technical skills and organizational and financial capacity to get involved, often coupled with burdensome regulatory provisions or inappropriate administrative structures and processes. For example, there are real risks that responsibility and authority will be applied to a 'community' level that is inappropriate in terms of what is known about effective collective action. Local groups may also have difficulty negotiating fair market prices, finding affordable transport, arranging payments, ensuring quality standards and meeting pre-payment requirements.

Ensuring best value for money

An important factor in weighing value for money is the intrinsic profitability of commercial forestry. If it is lower than the usual rate expected from long-term investments, the interest of the private sector will be small and may be only from those wishing to asset-strip with a quick exit. The government may have to promise to keep paying subsidies, promises that are usually discounted by the purchaser. The best approach will then be to seek out potential owners with a direct interest in wood production and other plantation products and services. The bidding process can be an effective and efficient means of getting the best deal.

Enforcing standards after transfer

Environmental standards in relation to outsourcing can be established and enforced through contracts; but this is not without its problems. Governments generally rely on the broad governance system to enforce environmental standards after transfer of assets. Standards may also be specified in transfer agreements. However, the costs to government of additional monitoring and enforcement can be substantial. Audit by a third party, with the costs borne by the beneficiary, may be an efficient alternative, though the transfer price will discount the additional cost.

Safeguards – for example, to maintain provision for public access – can be built into title deeds. The transfer price will reflect the cost to the new owner of such restrictions on their rights. It may also be possible to give first option to purchase forests with high nature conservation, recreation, cultural or heritage values to organizations mandated by their statutes to manage for the public good. Governments may choose to pay for environmental services; but the costs can be substantial, especially wherever higher outputs of public goods and services are demanded. Emerging markets for environmental services show some potential for ensuring continued provision of public goods more efficiently than other mechanisms.

Instruments for ensuring that private managers continue to serve the public interest after transfer are summarized in Box ES.1.

Managing social impacts of change

Change aimed at achieving greater efficiency, as with outsourcing, and transfers of state assets to more efficient owners and managers will be likely to lead to job losses from the sector. Those who have to leave the sector may find employment in other sectors so that, viewed from a perspective of national social welfare, there could be a net gain after taking into account the greater efficiency and profitability of plantations and the recruitment to other profitable enterprises outside forestry; but this will not always be the case. Those who remain in the sector may be faced with poorer conditions of service.

Box ES.1 Instruments for ensuring new owners/managers continue to serve the public interest

Obligations written into the transfer agreements

Obligations include:

- continuation of customary rights;
- South African model of retaining ownership in trust for land claimants;
- best management practice;
- performance indicators;
- certification as evidence of good performance; other information disclosure requirements;
- equity stakes;
- rental payments/benefit-sharing arrangements; performance bonds;
- duration and time-bound commitments.

Instruments external to the transaction

Instruments include:

- national and local strategic fora (e.g. national forestry programme and local governance fora), and other facilities for stakeholder dialogue;
- land reform, wider privatization programmes, political decentralization;
- laws, bylaws, regulation, constitutional guarantees, conventions;
- certification;
- taxation;
- payments for environmental services;
- subsidies for management and operations; reform of 'perverse' subsidies;
- information generation, access, brokerage and management;
- public information/awareness and extension actions;
- Millennium Development Goals;
- other sustainable development principles (becoming established in international law): the precautionary principle; the polluter pays principle; user pays; inter-generational equity; intra-generational equity; free, prior and informed consent of groups to changes such as development plans; and helping (involuntary) risk-bearers to participate in decisions, as well as risk-takers (government, investors).

Balancing tenure rights for bidders with other objectives

In partial or total asset transfers, beneficiaries need secure tenure rights. Weak tenure rights will lead to discounting of the price offered, or in the worst case no offers being received. However, outright tenure for one may be injustice for another. One of the most challenging aspects of the South African privatization process has been to secure reasonable tenure security while protecting communities' underlying land and other informal tenure rights.

Making change work

The context within which change is planned is all important; what works in one context will never quite fit the bill in another. History and power structure, the nature of the plantation asset base, ecological influences and constraints, economic and financial conditions, social–cultural influences and conflicts, institutional norms and precedents all influence the scope and rate of change. The forest governance system is critical; if it does not already have safeguards built in to protect against unwanted outcomes from change, it may have to be adapted before proceeding with privatization. Tactics for effective management of plantation ownership and management change are summarized in Box ES.2.

Bolstering the forest governance foundations

Land and property *tenure* needs to be secure and clear, and access to the courts must be available on equitable terms. Effective *regulation* is a prerequisite for achieving a healthy balance between private-sector investment and public needs. Key attributes of an effective regulatory framework include minimum employment conditions, penalties for damaging environmental impacts, and safeguarding valuable wildlife, cultural, heritage or landscape features. Government *institutions* need to be strong, smart and joined up. Government interventions need to be coordinated, even though the functions of institutions may differ. Policy and regulatory functions need to be separate from forest management functions in order to ensure clarity of responsibility inside and outside the government body.

Financial incentives, where they exist, need to be designed to deliver public policy objectives and avoid perverse outcomes. *Tax breaks* are especially difficult to design for the public good. *Payments* linked to the production of public goods are more transparent and more effective when a balanced mix of outputs is sought. Voluntary *certification* can work for the public good where there is strong market demand, where it is supported by government and where there is consensus on standards; where there are many small growers, mechanisms must be in place to enable cost-sharing and co-operative marketing.

Emerging *markets for social and environmental services* may secure some public goods, but could work against the interests of certain groups. They are not a 'silver bullet'. Different commodities work in different contexts. Competitiveness is difficult to achieve in nascent markets and governance is critical. There are implications for the distribution of costs and benefits; the livelihoods of poor communities may be threatened by the market through increased exclusion, lower incomes and a weaker asset base.

Setting objectives for change

Sharpen up your argument. Recognize that other actors have different values, encourage transparency and confidence in presenting them and negotiate practical objectives.

Negotiate a clear definition of 'the public interest'. Which goods and services provided by state plantations are threatened and who loses by changing ownership and management?

Keep objectives clear and simple. Potential opponents are more likely to buy into change if the purpose is clear and they can see beneficial outcomes.

Selecting and shaping the transfer option

Develop criteria based on sustainable development. Develop amongst actors a set of criteria based on fundamental elements of sustainable development and in line with national societal priorities by which decisions about ownership and management transfers can be judged.

Do your homework. Analyse the existing information base and carry out research to examine the options that may deliver the benefits that you want.

Plan for an optimum balance of powers. Aim to transfer all the rights that private-sector actors need to achieve optimum sustainability objectives and to ensure that government retains the rights necessary to achieve public-policy objectives.

Be prepared for trade-offs. Several aims may sit together but are likely to need reconciliation and compromise (e.g. attracting large-scale investment and encouraging small enterprise development).

Recognize that the transaction costs involved are high (e.g. in terms of the time required of officials in key ministries) even if, in the case of sales to the private sector, there is a realizable sale price.

Making it attractive and accessible to the target groups

Make sure that the resource is in good condition and free from fundamental conflict. To be of interest to investors and/or communities, resource quality and potential will be a critical determinant, as will the existence of challenges to land use for forestry.

Ensure transparency of process. Making a transfer process attractive to private-sector investors, community groups and government departments will require clear signals about who will do what and how they will be held accountable.

Build in sufficient security over use rights to encourage investment. Such security is likely to be a function of provisions in a lease or ownership agreement, including duration; the right to assign, sublet and mortgage use rights; the support of the transfer by a broader enabling policy and support services derived, for example, from land and institutional reform; and clarity over the ultimate ownership of the land/resource in the case of defined-period leasing.

Contract over a long enough period for the security and planning horizons of contractors and tenants. A guaranteed minimum tenure on entering the lease is crucial, with provision for early termination in the event of a material and un-remedied breach of lease conditions. A key issue is the lessee's confidence in the government's ability to deal with breaches of the lease's terms or the law.

Allow contract transfers. Making the lease assignable/transferable (in whole or in part) to another party makes use rights tradable. An assignable lease has a financial value best protected by practising sound management of the forest. Risks that use rights may be assigned to another, perhaps non-target, group to realize a quick profit need to be mitigated by requiring government's prior approval of the transfer.

Package services, assets or use rights in a way that will attract the target groups. This is best shaped through dialogue between the actors.

Address unfavourable investment climates (e.g. stemming from high taxation, remoteness to markets, expensive finance, over-weighty bureaucracy and adverse labour relations/costs).

Preparing the target groups

Estimate capacity. Make an assessment of the capacity of private enterprises to engage profitably with the transaction process and to meet their obligations; make full use of any rights that are transferred.

Promote continuous improvement of management systems. Encourage enterprises to develop, upgrade and continuously improve their systems for information generation and management; human resource development; participation; planning and management; finance management; and monitoring.

Support preparedness in community organizations. Community organizations often need support in addressing some of the following challenges:

- Generating trust among the actors and confidence that others will comply with agreements made.
- Building on existing forms of community organization rather than artificially constructed or administratively convenient units.
- Avoiding fragmentation with a large number of owners not bound by an umbrella organization or association.
- Ensuring complementarity of plantation and social units; collective action for resource management is more likely to occur when the boundaries of the resource and the boundaries of the social unit managing the resource coincide.

- Ensuring adequate financing of community management activities.
- Generating sufficient knowledge and expertise about plantation management.
- Overcoming conflict within and between community groups.
- Managing the long timeframes involved in tree-growing and occasionally the disincentives of seasonality clashes between farming and forestry activities.

Box ES.2 Tactics for effective management of plantation ownership and management change

Political buy-in

- Employ expertise in policy research, institutional development and good governance.
- Strengthen relations between decision-developers and the ultimate decision-takers.
- Work with the media.

Buy-in by civil society and private business

- Clarify, utilize and build on the complementary skills of the private sector and civil society.
- Use issues-based interactions.
- Deal with unrealistic expectations.

Capacity to manage the process effectively

- Maintain access to technical expertise.
- Secure a dedicated budget against a reasonably flexible timetable.
- Keep working to build effective capacity for realigned state roles, adequate private sector capacity and competition, as well as contract specifications that are based on quality management standards.

Managing the government's post-transaction responsibilities

- Clarify and communicate the powers and duties of the realigned regulating/supervisory/monitoring government.
- Build motivation to exercise these responsibilities.
- Provide ongoing support to enable private enterprise to manage plantations.
- Provide ongoing support to community-based plantation management.

Getting the best deal

Generate and commit to some principles for optimal deals. Whether transfers are planned to private enterprises or community organizations, commitment to some principles of good deal-making will help to produce an effective and equitable result.

Bid competitively. In the case of transfers to private enterprises, an open-market bidding-based approach to the transaction is crucial. Such auctions can allocate forest land use rights to the most efficient producer and can maximize sale price and revenue for the contracting authority. However, it is important to recognize that maximization of revenue comes at the expense of public policy objectives because they are a cost to bidders. Thus, sale price and revenue maximization may not be the highest objective.

Spread the word. It is vital that clear information about the resource and the proposed transaction process is developed and presented so that the target groups can access and digest.

Provide clarity on risks. Risks and burdens in the transaction will be dealt with by the private sector simply in terms of the price it is willing to pay. The greater the risks, the larger the discounting in the price offered for the asset. If the objectives of the transfer and transaction costs of the process are unclear, the price offered will also fall.

Recognize that requiring private-sector creativity to meet public-policy objectives increases risk. A competitive bidding process does not fully proscribe how investors should manage public policy issues; rather, it invites them to use their initiative in responding to them. But it is important not to overburden the transaction with so many public-policy objectives that it becomes unattractive to investors. Objectives such as revitalizing the plantation resource, investing in processing, and maximizing local ownership and employment may all present significant risks to the private sector and will have to be carefully weighed up.

Design tender systems for optimized objectives. The tender systems needs to be designed to enable selection of the bidder whose bid best reflects multiple objectives. This need not be the bid with the highest price. Qualitative criteria, such as commitments to future investment and opportunities for local participation and economic empowerment, need to be combined with quantitative criteria, such as the price consideration.

Evaluate bids against agreed criteria and each other. Potential investors are invited to compete against each other in response to the agreed criteria by submitting proposals, which might typically include a business plan and an offer price. Bids are then evaluated against the agreed objectives and each other to identify a preferred investor.

Conclusion

Transferring ownership of plantations can be the right thing to do when it puts power in the hands of those who can use plantations for equitable, efficient and sustainable ends. However, transfers can also go astray and be used to concentrate plantation power and privilege in too few hands. Absolute clarity of purpose, dedication of practitioners, specific steps, a phased learning approach and adequate resources, skills and time are all needed to make transfers work.

Acknowledgements

This book started life as a set of case studies and a synthesis paper prepared for the International Conference on Changing Ownership and Management of State Forest Plantations hosted by the Department of Water Affairs and Forestry (DWAF), South Africa, Department for International Development (DFID), UK, and the Food and Agriculture Organization of the United Nations (FAO) in Cape Town in November 2002. The conference brought together senior government officials, advisers and researchers from countries that had experienced – and were still experiencing – changes in ownership and management, and from countries where change was on the policy agenda. The International Institute for Environment and Development (IIED) was asked to steer the development of the case studies and the synthesis, to incorporate the contributions made by those who participated in the conference, and to reach a wider audience with the results. This books aims to do that.

We would like to thank our colleague on the synthesis, Natasha Landell-Mills of OTP Fund Management in Hungary, and the case study authors for their substantial contributions: Jacki Schirmer and Peter Kanowski of the Australian National University, Canberra; David Grundy (former commissioner of the UK Forestry Commission); Eduardo Morales of Fundacion Chile; Xu Jintao of the Centre for Agricultural Policy Research, Chinese Academy of Science, Beijing; William Hyde of the Centre for International Forestry Research, Bogor, Indonesia, and the University of Goteborg, Sweden; Sushil Saigal of Winrock International, India; Maude Dlomo, Chief Director, Education Human Resource Planning, National Department of Education, Pretoria; Mike Pitcher, formerly of Huntings Technical Services, UK; and Michael Roche, professor of geography, head of the School of People Environment and Planning, Massey University, Palmerston North New Zealand.

In their turn, the case study authors acknowledge the advice and comments they received and, in particular: for the Australia chapter, the many colleagues in and associated with Australian forestry who contributed information to and reviewed the chapter; for New Zealand, George Asher, Peter Berg, Eckerhart Brockerhoff, David Buckleigh, Peter Gorman, Bryce Heard, Murray Horton, Steve Jacobi, Edwin Jansen, Roger Keey, Lisa Langer, Hamish McDonald, Andrew McEwen, Rob McLagan, Devon McLean, Dennis Neilson, Colin O'Loughlin, Tony Pearce, Grant Rosoman, Guy Salmon, Charlie Schell,

John Stulen and Geoff Thorp; and for the UK the author's one-time colleagues in the Forestry Commission for their many illuminating comments.

For their inputs in the various stages of the organization of this work, analysis of the results and preparation of this book, we would like to thank Kimberly Clarke, Tim Foy, Julian Harlow, John Hudson, Merilio Morell, Elaine Morrison, Linda Mossop, Anelle Odendaal, Jake Paul and Eunice Shankland. We also thank the supporters of the Cape Town conference: DWAF, FAO and DFID.

The book was made possible by the generous support of DFID who paid for the research, writing and publication. Errors and omissions are the responsibility of the book's editors and chapter authors and the opinions expressed are their own.

Michael Garforth and James Mayers
July 2004

Introduction

Aim of this book

This book aims to improve understanding of the options for ownership and management of state forest plantations, their strengths and weaknesses in different settings, and how to generate the benefits of increased private-sector involvement without incurring unacceptable social, economic and environmental costs.

The book is about transfer of ownership, use rights, and powers and responsibilities for plantations from state organizations to non-state actors, such as private companies and communities. It is not about decentralization through layers of government to local administrations, which may be done to achieve some of the same purposes (and which frequently amounts only to transfer of responsibilities, not to transfer of power – see Box I.1).

We have tried to reflect the strength of differing opinion and controversy that can be generated by initiatives to manage transfers of ownership and control of state-owned plantations, and we have strived for a balanced critique of the actual processes and mechanisms involved.

The book has been developed from an overview paper prepared for the International Conference on Changing Ownership and Management of State Forest Plantations held in Cape Town in November 2002, and from seven country case studies (six of which were also prepared for the conference) and a review of literature. The case study countries are Australia (prepared by Jacki Schirmer and Peter Kanowski), Chile (prepared by Eduardo Morales), China (prepared by Xu Jintao and William Hyde), India (prepared by Sushil Saigal), South Africa (prepared by Maude Dlomo and Mike Pitcher), New Zealand (prepared by Jacki Schirmer and Michael Roche), and the UK (prepared by David Grundy). The case studies have been edited and brought together in this single volume.

Why this book is needed

Many people in many countries are grappling with issues linked to changing ownership and management of state-owned plantations; yet, there is a paucity of digestible lessons from experience and useful guidance.

The role of the state is changing in all sectors of the economy. Private-sector participation in forest ownership, management and operations has been increasing (Landell-Mills and Ford, 1999). Substantial areas of forest, including plantations, have been transferred from state to private ownership or use in some form during the last ten years. A large proportion of state-owned timber harvesting and processing has been sold to private enterprises and outsourcing of service functions is becoming the norm in many countries. Plantations may be seen as a softer transfer subject because they are perceived to lack the environmental and social capital associated with natural forests.[1] Experience shows that transferring ownership and management of plantations to private actors raises many of the same concerns that apply to natural forests, depending upon their characteristics and setting.

There are different motivations for changing from state ownership and management of forest plantations, including political philosophy, restoration of expropriated rights, increased efficiency of resource use, reduction in public spending, enthusiasm for vertical integration and investment in processing, the desire to attract foreign investment, stimulation of entrepreneurship to exploit opportunities, and improvement of rural livelihoods. There are also many different models.

Changing ownership and management raises a number of significant concerns – for example, increased corporate control leading to decreased opportunities for community decision-making or public input into forest management; loss of traditional rights; redistribution of plantation goods and services, leading to some people losing while others gain; and lower outputs and loss of non-marketed but valued services, such as landscape beauty and watershed protection services.

The choice of the most appropriate transfer option and of an effective change process is crucial if these and other concerns are to be dealt with and intractable problems prevented. This book identifies 'best-bet' practical instruments and processes for achieving increased sustainable private-sector participation, while guaranteeing the public interest, drawing from experience from around the world. Although it focuses on plantations, it has wider application. Many of the lessons are relevant to changing the ownership and management of forests of natural origin, a growing worldwide trend, with particular impact in some areas, such as Eastern Europe.

Structure of the book

The book is into two parts. *Part 1* draws from the experience of the seven countries studied to provide a overview of the opportunities to increase social welfare and economic performance by transferring ownership and management of state plantations to private actors, the challenges that transfers bring, and the principles and procedures that can help achieve successful transfers.

Box 1.1 Working definitions of some key terms

Forest: The United Nations Food and Agriculture Organization (FAO) adopted the following definition for their Global Forest Resources Assessment 2000: 'Land under natural forest or plantation with a minimum 10 per cent crown cover, excluding stands of trees established for agricultural production' (FAO, 2001a).

Plantation: The FAO adopted the following definition for their Global Forest Resources Assessment 2000: 'Forest stands established by planting or/and seeding in the process of afforestation or reafforestation. They are either of introduced or indigenous species which meet a minimum area requirement of 0.5ha tree crown cover of at least 5 per cent of the land cover and total height of the adult trees of at least 5m' (FAO, 2001a).

Private sector: The part of a country's economy owned and operated by private (non-governmental) entities (person, household, community, company) who seek to maximize their own benefits (from forest-related activities).

There are a number of key types of change and transfer of ownership, management and/or control of land and resources (Mayers and Bass, 1999):

*   *Decentralization:* the relocation of administrative functions away from a central location. Decentralization can be thought of as an overarching term for five different types of power (decision-influencing) transfer: deconcentration, delegation, deregulation, devolution or privatization.
*   *Deconcentration:* spreading authority from the central administration to its agencies closer to the 'grass roots'. A non-definitive transfer of decision-making and executive powers within the administrative or technical structure (e.g. from the Ministry of Interior to a governorship or from the national directorate of a service to the regional directorate). This takes the form of institutional modification from within an administration.
*   *Delegation:* a non-definitive transfer of authority from an administrative service to a semi-public or private company.
*   *Deregulation:* a transition in which a sector of activity previously regulated by a public authority ceases to be subject to such regulation.
*   *Devolution:* a transfer of power from a larger to a smaller jurisdiction; this transfer may be total or partial (e.g. transfer to local communities of decision-making over renewable resources on their village lands).
*   *Privatization:* a type of delegation involving transfer of rights, functions, obligations and responsibilities from the public sector to private entities, either directly or through parastatal institutions (*corporatization*).

Chapter 1 reviews the contribution that plantations make to sustainable development, the changes in ownership and management that have been taking place, and the conflicts that can arise between increased private-sector participation and protection of the public good.

Chapter 2 presents experience of plantation development and changing ownership and management from the seven case studies.

Chapter 3 draws lessons from the case study countries on how to reconcile increased private-sector experience participation with the public good, and presents practical guidance on how to achieved desired outcomes and avoid the possible pitfalls.

Chapter 4 draws conclusions on the extent to which the benefits sought from transfers have been achieved, identifies stumbling blocks to successful change processes, and suggests what needs to be done to ensure that initiatives to change ownership and management of state-owned plantations keep improving.

Those readers whose main interest is practical advice on how best to go about transferring ownership and management of state plantations to private actors should read Chapter 3 first.

Part 2 of the book presents the individual case studies in Chapters 5 to 11.

Part 1

Opportunities, Challenges and Solutions: An Overview

1

Plantations, Livelihoods and Poverty

Mike Garforth, Natasha Landell-Mills and
James Mayers

According to the Food and Agriculture Organization of the United Nations (FAO's) Global Forest Resources Assessment 2000 (FAO, 2001a), the total global area of plantations in 2000 was 187 million hectares: an increase from 17.8 million hectares in 1980 and 43.6 million hectares in 1990. The largest plantation forest resources are found in China (24 per cent of the world's total) and India (18 per cent). New plantations are being added at the rate of 4.5 million hectares a year, over 90 per cent of the increase being in Asia and South America. In contrast, the area of natural forest – 3682 million hectares – had declined at a rate of 16.1[1] million hectares a year through the 1990s: 1.5 million hectares of this loss being due to conversion to plantation.

It is important to note that the FAO definition of plantations does not include forests that are of natural origin but which, as a result of intervention, have lost many of the original forest's characteristics and whose non-timber values may be closer to plantations than natural forests. It also needs to be stressed that the FAO definition of plantations covers a wide range, from planting on farms (for example, as part of agroforestry systems) to industrial plantations. This chapter focuses on the latter type of plantations that have been the focus of privatization and restructuring efforts during the last 10 to 20 years.

Objectives for, and characteristics of, plantations vary: at one end of the spectrum a single rotation crop of trees with management emphasis on maximizing production; at the other, restoring environmental capital by increasing the permanent forest estate, with plantations a first step towards new forests that have many of the characteristics of managed natural forests (Sargent, 1992; Emborg and Larsen, 1999). Examples from across this spectrum can be found in many countries. As we shall see, objectives for plantations can change with time as governments, owners and managers respond to changing civil society needs and aspirations, international trading conditions and other factors.

Plantation goods and services

Plantations have the potential to produce many of the goods and services that natural forests are able to provide; the potential of plantations to contribute to sustainable development goals is recognized in intergovernmental agreements and programmes of action on forests (UN 1992a, 1992b), and by non-governmental organizations (NGOs) and private business (Forest Stewardship Council, 2000). There are some important exceptions, however. In terms of their biodiversity, plantations will never be as species or gene rich as undisturbed natural forests. The contribution of plantations to landscape beauty is contested, and cultural and heritage values of new plantations will not develop for generations, if at all.

Globally, FAO estimated that 48 per cent of plantations are for industrial use and 26 per cent for non-industrial use (FAO, 2001a).[2] The variety and volume of goods and services that plantations provide depend upon their characteristics and setting. Their characteristics are determined by owners' and managers' objectives and management practices, which are influenced by markets, government regulation, fiscal incentives and shareholders. Their setting may make them particularly valuable for conserving soil or for meeting the daily needs of local communities, for recreation, or valuable only for supplying material to wood-using industries.

Wood

Plantations provide *industrial wood* and pulp fibre; but they are also important in some settings for wood for *community and household use*, including for local *construction, fuelwood* and *charcoal*. In some countries, plantation-grown wood is being using increasingly as a renewable *alternative to fossil fuels*.

Although plantation forests account for less than 5 per cent of the world's forest cover, they account for at least 22 per cent of global roundwood supplies to industry. In several countries, industrial wood production from plantation forests meets a substantial proportion of industrial roundwood needs – for example, 99 per cent in New Zealand and 84 per cent in Chile (FAO, 2001a).

Forests are particularly important in developing countries for supplying wood for fuel, providing about 15 per cent of their total energy demand (WEC, 1999, cited in FAO, 2001a). Wood fuel provides about 7 per cent of energy demand for the world as a whole and industrialized countries only 2 per cent (FAO, 2001a). Production of wood fuel from plantations alone currently makes only a small contribution to energy requirements, although it is very important in some localities and countries. FAO estimated that plantations supply only 5 per cent of wood fuel, though production is likely to double by 2020 (FAO, 2001a). In developing countries, about one third of the total wood plantation estate was primarily grown for wood fuel in 1995 (FAO, 2001c).

Non-timber forest products

Some plantations are created for major non-timber forest products (NTFPs), such as black wattle in South Africa, which produces tannin used in the leather industry. Other plantation trees have important by-products for local livelihoods, such as pine resin. Plantations can sometimes provide important NTFPs derived from game or plants growing in association with the planted trees. Many animals and plants are important sources of food and medicine. In some plantations, food crop production is carried out alongside tree production in certain areas or tree growth stages, such as women's ground-nut growing groups in South Africa.

Carbon sequestration

When they grow, trees absorb carbon from the atmosphere. Forests thus represent important stores of carbon, accounting for about two-thirds of total terrestrial carbon. As a result, they play a critical role in the world's carbon cycle and are viewed by many as a key component of global efforts to reduce atmospheric carbon blamed for global warming (Bass et al, 2000; IPCC, 2000; Keenan and Grant, 2000). Creating new plantations could play a small but significant role in managing atmospheric carbon levels. The adoption of the Kyoto Protocol in 1997 triggered a strong increase in investment in plantations as carbon sinks (FAO, 2001g).

Biodiversity

In addition to housing valuable stocks of genetic diversity in their vegetation, plantations can provide habitats for a wide variety of wildlife. The level and value of biodiversity held in individual plantations depends upon their design and subsequent management. Structural diversity and the variety of tree and shrub species can be engineered to produce niches found in natural forests.

Enhancing biodiversity is one of the main objectives of plantation establishment in some developed countries that can afford to invest in restoring the environmental capital that they have lost as a consequence of their development (Forestry Commission, 1998).

Visual and recreational landscape

Plantations are used in many countries to improve urban and rural landscapes. Tree-planting is often an important component of land reclamation after mining operations – for example, former coal mines and gravel quarries in the UK and coastal dune systems in South Africa. Creating woodland strips along transport corridors reduces the visual impact of major highways. New woodland in and around towns enhances the visual landscape and provides recreational opportunities, contributing to improving the physical and spiritual well-being of local people. Woodland for recreation is especially important in developed countries with a high demand for outdoor recreation; for example, in 1998 there were over 350 million day leisure visits to UK forests, the majority of which are plantations (Forestry Commission, 2002).

Soil and water protection

Forests are widely credited with a number of soil and water benefits. These include the protection of steep slopes from erosion, reduction in soil salinity, maintenance of water quality, reduction in flash flooding and the protection of dry-season stream flows. The protection of these many benefits has been an important driver behind plantation establishment in critical catchment areas.[3] In China, the potential of forest plantations to combat soil erosion and poor water quality was one of the drivers behind the central government establishing a Forest Environmental Benefit Compensation Fund to pay for forest restoration (Lu Wenming et al, 2002). However, plantations are also blamed for the degradation of aquatic environments. In the UK, large-scale afforestation of some catchments has led to increased acidity of lakes as a result of the trees capturing acid emissions from power stations and factories. Poor plantation design has led to heavy shading of rivers and erosion.

Preservation of natural forests

It is often argued that plantations reduce logging pressure on natural forests. If plantation development is targeted at the most appropriate ecological zones, and if sustainable forest management principles are applied, forest plantations can provide a critical substitute for natural forest raw-material supply (FAO, 2001a).

Economy, livelihoods and poverty

Contribution to national economies

Globally, in absolute and percentage terms, plantations contribute a tiny fraction of gross domestic product (GDP); but their contribution is significant in countries that have developed large areas of forest plantations and that have comparative advantage owing to high growth rates, low costs of land and low unit or gross labour costs.[4] New Zealand's plantation-based forestry industry, for example, has been estimated to contribute 4 per cent of GDP compared with the UK's 0.15 per cent. Favourable policy and silvicultural environments can attract foreign investment in plantation development and management, wood processing and in the non-timber capital of plantations (e.g. carbon). International sales of plantation products can contribute to export earnings. Box 1.1 illustrates the contribution that plantations are making in two

Box 1.1 Plantations' contribution to two national economies: South Africa and New Zealand

South Africa

South Africa's forest products industries – particularly the pulp and paper component – have grown rapidly. From an effective base rate of zero during the 1940s, the pulp and paper industry now produces around 2.8 million tonnes of pulp (1.63 per cent of global supply), 2 million tonnes of paper (0.76 per cent of global supply) and 1.3 million tonnes of sawn timber (0.3 per cent of world production). South Africa has moved during this period from being a net importer to a net exporter. In terms of its contribution to the South African economy with a gross domestic product (GDP) of approximately US$110 billion, the industry generates in excess of US$1 billion, which:

*   Meets 90 per cent of domestic forest products needs.
*   Contributes about 1.5 per cent to GDP.
*   Contributes approximately 5.5 per cent to the manufacturing sector's contribution to GDP.
*   Provides 4.7 per cent of total export earnings and about 10 per cent of manufactured exports.
*   Employs approximately 110,000 people.

The latest available information (DWAF, 2003) indicates that forestry contributes 8.5 per cent (2091 million rand) of agricultural contribution to GDP, and forest products represent 7 per cent (9144 million rand) of manufacturing GDP. Together, forestry and the associated processing of forest products generate 1.5 per cent of South Africa's GDP. This belies the fact that plantation forestry often operates on poorer sites not suited to high-return agriculture. Despite this, even without considering the value-adding processing, forestry returns twice the value per hectare than the average for agriculture (a contribution of 1571 rand per hectare compared to the average of 751 rand per hectare). If one adds processing of forest products, forestry contributes 8441 rand per hectre to GDP.

New Zealand

The contribution of the forest industries – which are predominantly made up of the plantation industry – to the national economy during 2001–2002 was estimated at 4 per cent of national GDP (NZFOA, 2002). It was estimated that the forestry and first-stage processing industries directly employed 24,315 people in February 2001 (NZFOA, 2002). Employment opportunities are forecast to increase, and there have been recent shortages in labour supply for forest growing and harvesting (MAF, 2002a).

Approximately 13 million cubic metres were exported in raw and processed form, bringing in NZ$3.7 billion to the domestic economy (NZFOA, 2002). The harvest from plantations in New Zealand has increased steadily and is forecast to keep increasing, with a predicted annual harvest of 41.9 million cubic metres by 2025, and 52.5 million cubic metres by 2040 (MAF, 2002a).

The domestic forestry industry has enough plantation resources to absorb NZ$3 billion more in investment in wood processing capacity, based on the assumption that 25–30 per cent of logs are exported. Between 1995 and 2002, approximately NZ$1 billion in planned wood-processing investments were announced, leaving a large gap between future available wood resources and domestic processing capacity (MAF, 2002a).

countries – South Africa and New Zealand – which have pursued aggressive plantation development policies over many decades.

Rural livelihoods

The contribution of forest plantations to local and national economies is far higher than GDP estimates for the sector would show. Livelihoods of rural people are enhanced by employment in large-scale plantings or participation in out-grower schemes (FAO, 2001f). Out-grower schemes are virtually standard in the pulp sector – 60 per cent of pulp-producing companies covered in a global survey source some of their product from out-growers (IIED, 1996). Depending on tenure and access rights, rural people may also be able to benefit from using or selling timber and non-timber products or by selling plantation-based services, such as recreation, tourism, carbon or even watershed protection services (Landell-Mills and Porras, 2002). Plantation development may also create important social infrastructure associated with commercial forestry, including roads, housing and, in some cases, schools and clinics.

In South Africa, plantations provide the basis for a wide range of small-scale processing and retailing enterprises, including small-scale saw-milling, furniture manufacturing and fuelwood selling, and the commercial growing of indigenous trees for high-value medicinal products (see Chapter 10). The plantation forest and forest products industry employs over 150,000 people in full-time waged employment, of whom about half work in production forestry and half in wood processing industries. If supporting industries are included, the sector contributes about 600,000 jobs to the economy. Assuming a dependency ratio of five (i.e. five family members depend upon one income earner) then 3 million people (about 7 per cent of South Africa's population) depend upon the sector. Direct access to forest plantation products (e.g. building materials and fuelwood) and other resources (e.g. grazing) by rural households is important in some countries. Many households in developing countries depend upon wood for energy. As much as 20 per cent of South Africa's total energy consumption is derived from biomass, primarily woody biomass from the forest and woodland resource. In the state of KwaZulu Natal it is estimated that wood-fuel use has an annual value equivalent to provincial electrical use.

Forest products such as small timber, bamboo and grasses are extensively used in India for house construction and manufacturing agricultural implements. The consumption of bamboo alone for house construction was estimated at 1.6 million tonnes a year in the 1980s (GoI, 1984). A projected estimate puts the total demand for wood for construction at 11.5 million cubic metres (cum) a year (Rao, 1978, in Tewari, 1995). Tewari (1995) estimated that 2.1 million bullock carts, 50 million yokes, 100 million wooden ploughs and 30 million wooden seeders are constructed each year. Most of this demand is met by local artisans who use local raw materials and traditional skills.

Poverty

High numbers of rural poor live near plantations in some countries. Plantations may do little for the rural poor through local economies and may take up land more rightly suited to poor people's food production. However, in some rural areas plantations contribute to an effective pattern of local land use and provide some local benefits to the poor where few other economic opportunities are available. Out of 260 million poor people in India, over 75 per cent (197 million) reside in rural areas and constitute 27.1 per cent of the rural population (GoI, 2002a). Thirty-two million hectares of forest are distributed in approximately 170,000 villages, having a total population of about 147 million (FSI, 1999). Forest-based small-scale enterprises based on non-wood forest products provide up to 50 per cent of the income for 20 to 30 per cent of the rural labour force (GoI, 1999).

Evidence from out-grower plantation schemes suggests that significant and relatively equitable benefits can flow to poor people; but much investment in local bargaining power is needed before such schemes can demonstrate pathways out of poverty (Mayers and Vermeulen, 2002a). In South Africa, some 18,000 households are small-scale commercial timber producers, mostly in KwaZulu-Natal Province, with a total planted area of around 43,000 hectares (DWAF, 2003). Although accounting for less than 3 per cent of the national plantation resource, the economic contribution made by small growers in some rural communities is significant. With a national poverty line calculated at ZARand1100 rand per month, it is estimated that the out-grower schemes contribute, under average management, from 12 to 45 per cent of the income needed for a household to remain just above this line (the figure for 1 hectare is 17 per cent) (Mayers and Vermeulen, 2000a). Three conclusions can be drawn from this: firstly, such schemes can offer real contributions to rural livelihoods and the reduction of poverty; secondly, while poverty may be reduced, it is unlikely to be eliminated by such developments; thirdly, each household would need a minimum of 6 hectares to obtain a poverty line income. The potential of out-grower schemes may be limited by land tenure constraints and poor transport infrastructure. In South Africa, licensing and the national priority assigned to water management is an additional limitation.

The contribution of plantations to alleviating rural poverty depends crucially upon tenure and access rights, mobility of the workforce and the labour demand of the plantation. In Australia, the employment created by plantations is likely to be in regional centres rather than the small, remote rural communities that have the highest rates of unemployment and poverty. Some small rural communities have raised concerns that plantations established in their locality do not generate employment in their community, with employment created instead in large regional towns some distance away, and only irregular employment available. The amount and type of employment provided by plantations – whether private or state – depends upon the type of plantation being established and the level of associated downstream processing. The amount and type of employment (direct and downstream) provided by short rotation eucalyptus plantations established for wood chip production is likely to be lower than the amount provided by *Pinus radiata* plantations, which are pruned and thinned during their rotation and which are processed domestically into a wide range of products.

Changing ownership and management: Some global patterns

In countries that have developed significant forest plantation resources, governments have generally played two roles: as plantation developer, either by purchasing land and establishing plantations through state organizations or through joint ventures with landholders, and as sponsor of private plantation development through favourable policy environments and direct (grants) or indirect (favourable tax regime) support.

The FAO's Global Forest Resources Assessment 2000 estimated that, globally, industrial plantations were 34 per cent publicly owned and 29 per cent privately owned.[5] Within non-industrial plantations, 41 per cent were publicly owned and 37 per cent privately owned.[6] These figures hide large variations in the distribution of private-sector holdings between large, corporate industrial owners and smaller holdings owned by private individuals or community groups.

Private-sector participation in forest ownership, management and operations has been increasing. As governments around the world adopt more liberal market policies and are under pressure to reduce budget deficits, they are loosening their control over private-sector activity and are encouraging individuals, communities and firms to provide goods and services traditionally provided by government (Landell-Mills and Ford, 1999). Substantial areas of state forest, including plantations, have been transferred from state to private ownership or use in some form during the last ten years. A large proportion of state-owned timber harvesting and processing has been sold to private enterprises, and outsourcing of service functions is becoming the norm in many countries.

Models of private-sector involvement

Governments and state forest services have employed several models of private-sector involvement in state-owned forest plantations, often in combination:

*   *Outsourcing of services*: the state enterprise retains ownership and responsibility for deciding the flow and distribution of goods and services but outsources management and operational activities – for example, inventory, management planning, silvicultural operations and forest protection functions.
*   *Transfer of use rights* of one or more good or service: the state enterprise retains ownership of the land and may retain rights to decide the flow of goods and services. Rights may be transferred under short- or long-term agreements to harvesting companies, communities or households. Often rights are transferred alongside management obligations, which are defined in agreements between the 'landlord' and 'tenant'. Obligations may be to avoid negative impacts and/or to achieve certain outcomes.
*   *Transfer of ownership* of the right in perpetuity to use, sell or transfer to other users: the process may be to the highest bidder or to a preferred beneficiary. The special case of *restitution* transfers 'tenure, revenue ownership and management rights to private individuals or bodies corporate by way of handing back productive assets to their former owners' (Indufor and Eco, 2001). It has been estimated that restitution will create 3 to 4 million new private forest holdings in the ten countries that acceded to the European Union (EU) in 2004, with an average size of 2 to 3 hectares. Ensuring the profitability and sustainability of these new small holdings will be a major challenge for these countries (European Economic and Social Committee, 2002).

Limited consumption rights versus continuous decision-making rights

There is an important distinction between granting rights to consume specified forest goods and services within extraction limits set by the owner, and transferring rights to decide the flow and distribution of goods and services. An example of the first is the permission given to rural households in South Africa to take head loads and truck loads of timber out of the forest to use for fuel or local construction (see Chapter 10). The users in this example play no part in deciding the objectives of forest management, in planning and implementing harvesting and regeneration operations, or in protection and maintenance activities as part of the formal or informal agreement under which they are granted rights. They take on no obligations other than to avoid causing any damage to the forest. The users do not have exclusive rights; the landowner retains ownership of the product until it has been consumed and may use it themselves without infringing the rights of other users. In these cases, there is no change in the ownership and management of the assets held in the forest. Such cases are only considered tangentially in this chapter.

A privatization continuum

There is a continuum between the two extremes of unfettered state control over forest resources, forest management and forest utilization, and unrestricted individual ownership rights. Between these extremes there are varying degrees of leasing and sub-contracting (Indufor and Eco, 2001). Neither extreme will be achieved fully in reality; most European countries, for example, have chosen to preserve, through legal means, a mixture of ownership categories with public supervision over forest management, and enforcement of laws regulating the forest sector to protect forests against conversion and deterioration in quality and to ensure the supply of public goods and services that people demand from forests.

Opportunities associated with transferring ownership and management

Changes in ownership and management have been driven by different motivations. Increased operational efficiency and profitability of state-owned enterprises and reduced public spending are stated motivations, in most instances. Increased investment in wood processing by the private sector and a desire for increased export earnings have been significant drivers in a number of countries.

Righting past wrongs has driven the restitution of forests to private owners in Eastern Europe; similar motivations have influenced the change process in South Africa as the country seeks to realign its policies and institutions away from the inequalities and distortions of the apartheid era and towards a new paradigm in which much is expected of the private sector in serving society's needs.

Poverty reduction and improved rural livelihoods are targets for changing ownership and management in China, India and South Africa. Empowerment of previously disadvantaged groups has been a major influence on the design of divestment options in South Africa.

In some cases privatization, or moves towards it, has been driven by political philosophy, as in Chile, where it met with no resistance, and in the UK, where it was blocked by concerted opposition from NGOs.

Increased economic efficiency and improved aggregate welfare

Economists point to two major forms of efficiency: allocative and productive efficiency. Productive efficiency refers to the ability of the economic agent to maximize outputs with given inputs. Allocative efficiency refers to the optimal distribution of land/natural resources, labour and physical capital within an economic system; where resources are correctly allocated and productively employed, welfare will be maximized.

State intervention in the free working of markets is thought to undermine both forms of efficiency and, thus, human welfare. This is because governments fail in a number of ways:

- Firstly, their capacity to collect information on all of society's demands and on all costs of production, and to analyse this information to determine the optimal allocation of resources, is limited. Markets process this information automatically – Adam Smith's 'invisible hand'.
- Secondly, governments are not unified bodies, but are composed of individuals with their own interests and aims. There is no clear reason why this grouping of selected 'leaders' should be better placed than private individuals to determine production and consumption.
- Thirdly, and linked to the above, government officials are vulnerable to rent-seeking. Where competitive markets are distorted, outcomes are no longer determined by competition but by opaque decision-making processes that provide opportunities for corruption.

The market optimality argument is a powerful driver for privatization in the plantation sector. The private sector will usually have greater flexibility, better access to finance, fewer restrictions, more drive and clearer objectives. And the argument does not apply only to raising efficiency in tree-planting; plantation privatization may also be a lever for raising efficiency in the processing sector – for example, by increasing effective competition for raw material supplies.

Of course, there are exceptions to the market optimality argument. The existence of public goods and externalities, for instance, means that markets fail to operate effectively (see Box 1.2). Private-sector efficiency may conflict with public interests, such as continued use by local people and protection of wildlife. The challenge is, then, to weigh up whether government or market failure causes more damage, or whether an intermediate public–private or even community mix can offer a solution.

Controlled budget deficits
Where unprofitable plantation operations act as a drain on government resources, they may contribute to worsening budget deficits. Governments' losses in the arena of forest plantations have direct implications for resources available to other sectors, such as health and education. Often these trade-offs are not made explicit. They are, nevertheless, very real and significant.

Government budget deficits also have broader implications for welfare throughout the economy, translated through their impacts on interest rates and inflation. Budget deficits can be financed in two ways: by borrowing and by printing money. Heavy government borrowing can push up national interest rates, with negative implications for private borrowers. Often the first to be hurt by rising interest rates are the poorest. Where authorities choose instead to print money, the damage is done through rising inflation. Again, inflation often hits the most vulnerable hardest by devaluing their savings and reducing the real value of their income (as they tend to be least able to lobby for regular increases).

Box 1.2 Public and private benefits: Economic distinctions

Forest resources have been owned and controlled by national and state governments because of the belief that many forest goods and services would not be properly produced and allocated under a system of private ownership and market exchanges. Indeed, forests produce:

- goods that are well suited for market allocation and private consumption, such as timber;
- services that cannot be rationed by a market system and tend to be considered public goods, such as forest recreation, carbon storage and biodiversity.

Two key concepts help to distinguish between goods that are best suited for market allocation and goods which, due to market failure, are often considered to be public goods:

1 Excludability: where an individual can deny the use of any goods or service to another individual.
2 Sub-tractability: the amount that the consumption of a product or service subtracts from its sustainable consumption.

Most consumer goods, like timber, can only be consumed once – they are highly sub-tractable. And, since it is also easy to exclude other individuals from using consumer goods, these goods are thought to be most efficiently allocated by the market.

On the other hand, biodiversity is characterized by low excludability and low sub-tractability, and is then best treated as public goods, subject to governmental regulation. Since there is little incentive for an individual to invest in the provision of biodiversity, it will tend to be underprovided – or not provided – unless a government or an association accepts the responsibility for the provision for the public's benefit.

However, there is not always a clear-cut case for private or public control of specific goods and services. This is because the public or private nature of goods and services is not static, but depends upon the level of institutional sophistication, communications and technology. It is possible to change excludability and sub-tractability through a number of measures (e.g. zoning and management agreements). Hence, there is potential to transfer what once had to be public goods to market provision, with institutional improvements and appropriate safeguards. Moreover, public and private provisions are not the only options. Mixed public/private goods can be very effectively managed under strong common property regimes.

Forest Services	Goods and excludability	Subtractability	Externalities and comments
Timber	High	High	Private goods
Hunting	Medium	Medium	Private – congestion effects
Grazing	Medium	High	Mixed public–private
Fuelwood collection	Medium	High	Mixed public–private
Non-timber forest product (NTFP) collection	Medium	High	Mixed public–private
Recreation/amenity uses	Medium	Medium	Public–congestion effects
Carbon sequestration	Low	Low	Public goods
Micro-climate moderation	Low	Low	Public goods
Watershed protection	Low	Low	Public goods
Biodiversity conservation	Low	Low	Public goods

Source: Bass and Hearne, 1997

Gains to the public purse may be offset by the costs of ensuring that the new owners comply with forestry regulations and deliver public benefits that continue to be demanded by society after privatization. Such costs may include additional forest authority staff and payments to the new owners for environmental services.

Reduced state power and widening ownership
The greater the quantity of resources that states control, the greater their power over their citizens, and the greater their opportunities for corruption. Even where governments are democratically accountable, the machinery of the state can accumulate power in ways that were never intended, and systems for ensuring accountability are not always effective. Forestry officials that manage millions of dollars worth of plantation assets may not necessarily manage these resources in the national interest, or in the interests of local people. Privatization may offer a route for returning power to local people and for promoting a wider and more representative pattern of ownership.

Increased participation by disadvantaged groups
Privatization may be a useful means of increasing participation in the economy by previously disadvantaged groups (e.g. black empowerment in South Africa through direct participation in forestry operations; training and skills development; affirmative action in management; and entrepreneurial opportunities through outsourcing, partnerships, procurement and easier access to financing; see Chapter 10).

Increased entrepreneurial drive and investment

Privatization transfers rights and responsibilities from state actors to non-state entities. Where the private sector has traditionally been excluded from plantations, its energy and drive for improvements and technological advance would have been stifled. By encouraging private entrepreneurs back into forestry, the government also opens the door for increased innovation and longer-term growth.

Privatization offers opportunities to attract the investment and expertise needed to revitalize assets that are perceived to suffer from chronic under-investment. It is not always the case, however, that past investment by state agencies has suffered from a lack of resources, but instead from an institutional inability to spend the resources efficiently to achieve economic sustainability and growth. Dlomo and Pitcher (see Chapter 10) highlight this effect, citing the example of the homeland plantations in South Africa.

Removed contradictions between government as regulator and as manager

State management of plantations can conflict with the performance of governments' regulatory role. For example, wittingly or unwittingly, the commercial interests of the state agency may be put before the interests of private plantation owners competing in the same markets. Even where separation between the two roles avoids conflict of interest, the dual responsibility may reduce operational focus and, thus, operational effectiveness and efficiency. Transfer of plantation management to the private sector may improve operational focus on the regulatory role.

Rethinking land use

Privatization debates and negotiation processes may reveal the possibilities of alternative mixes of land use. For example, some plantation areas considered in South Africa's privatization process are now being managed for rehabilitation of natural habitat and other non-plantation uses.

Multiple motivations

Often there are multiple motivations. The driving forces behind New Zealand's 'corporatization' and subsequent sale of its state forest management enterprise are a good example (Clarke, 2000; Wijewardana, 2000). Wood supply from New Zealand's state plantations was forecast to surge during the 1990s; a more commercial operating environment was necessary to maximize returns. This required downstream investments, which were impeded by the government corporation's limited ability to raise capital – the shareholding ministers having to approve any investment intention – and by private-sector processors' reliance on contractual wood supply arrangements. The second major driver was the government's policy of clarifying organizational objectives, thereby enabling transparency and accountability. And the third driver was pressure on the government from the environmental movement to ensure more environmentally friendly forestry practices.

Private profit versus public good

Many fear the impacts of growing private investment in forestry. This concern stems from a belief that private companies are not accountable to public demands and have no incentive to provide important environmental and social goods and services. The private sector, it is argued, answers only to shareholders and their chief aim is profit.[7] The public sector, in contrast, is thought to be more accountable to the population at large (at least where governments are democratically elected). Moreover, because market failure means that private actors will not adequately supply public goods, it is up to the public sector to take the lead in sectors that suffer pervasive externalities. While the debate is often polarized, reality is far less clear cut; there is no absolute case for private or public control of specific goods and services.

The public or private nature of goods and services is not static but depends upon the level of institutional sophistication, communications and technology (Mayers and Bass, 1999). Whereas landscape beauty has traditionally been viewed as a public good, we increasingly see private parks offering unique views for sale. Clean air, also traditionally viewed as a public good, is already being commercialized in some quarters as international agreements under the United Nation's Framework Convention on Climate Change (UNFCCC) take shape. Even where public goods persist, whether or not the public sector offers a solution depends upon its capacity. The costs of market failure must always be weighed against those of government failure.

Furthermore, the dichotomy often presented between public and private is misleading. Mixed public/private goods can be very effectively managed under strong common property regimes. Hence, there is potential to transfer what once had to be public goods to market or community systems with institutional improvements and appropriate safeguards (Landell-Mills and Porras, 2002).

Public policy versus civil society aspirations

Changing ownership and management results in shifts in power to determine forest plantations' short- and long-term futures and to determine flows and distribution of benefits. These shifts may have a number of impacts that will lead to conflicts over the objectives of changing ownership and management between new owners/managers and wider public-policy objectives or civil society aspirations. The degree of conflict will depend upon the extent to which aspirations diverge, the value of the forest to different actors, and the power balance between the new owners/managers, forest regulatory bodies and civil society, which is itself determined by the governance system.

The governance system within which a change of ownership and management takes place varies greatly from country to country. Central government often has overarching powers, checked and balanced through intervention by local government, local groups, NGOs and private business, depending upon their respective capabilities and how much influence they are allowed or have the capacity to exercise. Conflicts can arise between the government's perception of the public interest, values important to civil society, and values important to owners and users to whom power and rights are transferred.

A country's governance system may limit the options and any substantial change in ownership and management may require substantial change in governance: re-engineering of the legal and institutional framework; decentralization of administration; giving capacity to new owners and managers; and strengthening new or changed government institutions.

Hostile public perceptions of transfers of ownership, particularly of privatization, should not be underestimated. In many countries, people perceive an over-hasty rush to allow protected public resources to be cracked open by the private sector and then fenced off by the market. They see mass privatization and deregulation as having bred armies of locked-out people, whose services are no longer needed and whose basic needs go unmet. The extent to which these perceptions are valid in the case of state-owned plantations, and the degree of engagement with such concerns, will be critical determinants of the success or failure of processes of transfer of ownership and control.

Key concerns

Specific conflicts around the issue of transferring ownership and management to private actors are associated with three main sets of concerns. Examples of these are provided below.

Economic concerns

Loss of timber and non-timber forest products. Concerns are often raised that privatization of government plantations will lead to resource 'mining' and the loss of valuable timber and non-timber forest assets as companies seek to recover the costs of their (often minimal) investment in as short a time period as possible.

Decline in processing capacity. Where the state has acted as the main supplier of raw materials in the forestry sector, it has often played a key role in supporting a fledging processing sector. In many cases, this support has been in the form of cheap logs, effectively acting as a subsidy to the industry. Where the government considers privatizing its plantations, the repercussions for downstream processing could be enormous and is thus often fiercely resisted by major industry players.

Social concerns

Restricted access. A major concern for communities living in or near plantation areas is that privatization will impact upon traditional use rights and livelihoods, or conflict with public recreational and spiritual pursuits. In some European countries public access restrictions protect landowners' rights to manage and use their land for their own goals, clearly limiting public opportunities to enjoy forest land (Jeanrenaud, 2001). In South Africa, serious questions have been raised about how the restructuring of government plantations can be carried out while preserving the land rights of dispossessed former owners (Clarke, 2000; see also Chapter 10).

Job-shedding. Where government-run plantations have been criticized for inefficiency and low profitability, privatization is often associated with a reduction in spare capacity and labour redundancies. The social costs of this process may be significant in forest-dependent communities.

Environmental concerns

Conversion of plantations to other uses. If plantation owners see value only in trees, there are risks that they will choose to replace their relatively longer-term riskier investment with shorter rotation alternatives, or simply 'cut and run'. Where conversion does occur, it is often argued that valuable environmental services are lost. The strength of this argument depends upon the characteristics and setting of the plantation. Whether new owners would replant and expand plantations was a controversial issue in the New Zealand privatization (Clarke, 2000). In South Africa, the privatization debate has raised the possibility of other land uses better suited to deliver environmental services (see Chapter 10).

Degradation as a result of overgrazing or neglect. Where privatization is associated with a downturn in the timber market, there are risks that plantations will be converted to other uses and that forest owners will choose not to invest in forest maintenance and protection. Again, plantation degradation is frequently blamed for lost environmental assets.

Destruction of biodiversity, landscape and watershed values. Even where private plantations are maintained, forest survey and management planning may not put as much weight as would local people and NGOs on the value of 'non-productive' assets, such as (amongst others) biodiversity, ancient monuments, ancient trees and the valuable watershed services delivered to local communities.

Increased pollution. This may be caused by 'cutting corners' to reduce costs (e.g. spillage or unsafe disposal of waste engine and lubricating oil).

Many of the above fears are expressed through opposition to transfers. Whether or not it is justified, in some cases opposition may be strong enough to block government plans, as in the UK (see Chapter 11). In other cases, concerns go unheeded and store up problems that emerge later. Therefore, when transferring ownership or management of state forest plantations to the private sector, it is critical that governments assess the many economic, social and environmental impacts their policy is likely to have. The transfer process needs to be tailored to minimize the costs and maximize the benefits.

The challenge of balancing risks and rewards for the private sector

As both the 'seller' of forest plantations and the regulatory authority, the government is responsible for guaranteeing the public interest in any initiative to privatize plantations. How can governments protect the public interest at the same time as securing the benefits sought from change of ownership and management? A range of instruments is available to governments: regulation; the design of the tenure system (e.g. the procedure for allocation, length of tenure and transferability of rights); financial incentives; extension; payments and markets for environmental services, such as carbon and biodiversity; and certification as a complement to, or substitute for, regulation.

The most effective balance between public and private ownership and management depends upon what people want. There are trade-offs. It may not be possible for everyone to win; but if some people feel they are losing, the balance may be wrong. The most equitable balance may not be the most efficient economically. Restitution has created problems in parts of Eastern Europe by breaking large enterprises into millions of small holdings. But righting the past wrong of expropriation of private land has been viewed as more important in those countries than economic efficiency in forest management and use.

The key question facing governments is whether they have the knowledge and capacity to ensure that the private sector makes good on its commitment to do the positive things expected to occur from privatization, thereby raising welfare while minimizing any negative repercussions.

Governments cannot rely only on the imposition of long lists of restrictions on private entities. There is a real risk that government infringements on private enterprise will chip away at incentives to reinvest in forest maintenance, thereby undermining the very public goods and services the laws have been written to protect. A number of examples illustrate this point (see Box 1.3). Private entities must be rewarded for the risks they take on when investing in what remains a long-term and uncertain business. Without prospects of commensurate rewards, no private investment will be forthcoming (Lu Wenming et al, 2002). While forest conservation for the protection of environmental services may be in the public's interest – for instance, where requirements for conservation undermine a plantation's production potential – few private entities are likely to invest (Cai, 1999).

Before moving forward with transfers – whether this is of ownership, use rights or management responsibilities – public authorities need to determine whether society's minimum requirements for public goods derived from plantations are compatible with private investment. If the two are deemed compatible, governments need to ensure that the private sector actually delivers the positive things expected from privatization. The government's success depends upon it finding a sustainable balance between stakeholders' needs. Where change concerns rights to determine the flow and distribution of goods and services through transfer of ownership or conclusion of long-term lease agreements, balancing acts need to be performed at two stages:

1   *During plantation transfer:* the balances attempted specifically to achieve optimum outcomes when transfers are made (e.g. restrictions or privileges built into transaction documents).

Box 1.3 Impacts of over-regulation

Indufor and Eco (2001) observed that in Western Europe and North America, changing political and economic conditions have led to a situation where individual private owner's rights to determine property management objectives and to exclude others from the use of the forest are being increasingly limited. In order to meet the public's rising demand for public benefits, 'social responsibilities' have been defined for private forest owners and imposed through legal and administrative action. Where owners go uncompensated, the chances that they will continue to invest in their forest plots lessen.

In China the imposition of restrictive harvesting quotas in 1986 went hand in hand with efforts to raise private-sector participation in tree-planting and management. While viewed as the central plank of the State Forestry Administration's system of control, the quota system is criticized for being excessively rigid and undermining private motivation (Lu Wenming et al, 2002). Xu and Hyde (2002) observe that government regulations on timber harvest levels and shipments remain a key deterrent to timber markets development in some regions.

Phoung (2000) emphasizes similar problems in Vietnam, where farmers still have to follow the instructions of the state management agencies with respect to species selection and business decisions.

2 *After plantation transfer*: the balances attempted on a day-to-day basis to make it work. These are rooted in the existing system of institutions, laws, regulations and incentives that enforce or encourage responsible plantation development and management (i.e. the forest governance system) (Mayers et al, 2002).

2

Plantation Development and Management: Drivers, Governance and Changing Roles

Mike Garforth, Natasha Landell-Mills and James Mayers

In this chapter we look at the case study countries' experience of plantation development, private-sector involvement in plantation resources created by the state and the governance framework within which private actors operate. The countries were selected for the diverse experiences they offer. They represent the full range of political, social, environmental and economic states and trends, the value of forests to national economies and local livelihoods, motivations and privatization models. The context for changing ownership and management in each of the countries is profiled in Table 2.1.

Drivers of plantation development

The seven case study countries have a total of 86 million hectares of plantation (46 per cent of world total), constituting 20.6 per cent of the countries' total forest estate (see Table 2.2). Two countries, India and China, have 90 per cent of the total (77.7 million hectares). The proportion of the countries' total forest estate composed of plantations ranges from 0.9 per cent in Australia to 67.9 per cent in the UK.

Plantation development has occurred mainly with government support, either through direct involvement in plantation establishment and management, or as a result of government-inspired or government-funded incentives for establishment by the private sector. There are exceptions – notably, the continuing expansion of plantations in New Zealand by private actors in spite of all subsidies having been removed, and also the early years of plantation development in Chile and small-scale plantings in the UK during the 19th century.

Table 2.1 *Context for changing ownership and management in the case study countries*

	Australia	Chile	China	India	New Zealand	South Africa	UK
Social and economic context							
Developing economy		✓	✓	✓			
Post-industrial economy	✓				✓	✓	✓
Large number/high proportion of rural poor			✓	✓			
Value of forests to national economies and livelihoods							
Relative to timber, high values placed on recreation, wildlife and visual landscape values of forests	✓				✓	✓	✓
Significant contribution to the national economy (>1% of GDP)		✓			✓	✓	
Main motivations for restructuring							
Philosophy of regulated market economy	✓	✓	✓		✓	✓	✓
Reduction in public debt	✓	✓			✓	✓	✓
Increased profitability/efficiency of the sector	✓	✓	✓		✓	✓	✓
Increased investment in downstream processing	✓				✓	✓	
Empowerment of disadvantaged groups				✓		✓	
Privatization models							
Transfer of the entire stock of state forests to private for-profit companies	✓	✓			✓	✓	
Limited asset sales	✓						✓
Transfer of use rights to individuals, households and private 'for-profit' companies			✓	✓			
Outsourcing of service provision to for-profit companies	✓		✓		✓	✓	✓
Outsourcing of service provision to communities and households	✓		✓	✓	✓	✓	✓
Outsourcing of service provision to not-for-profit agencies							

Table 2.2 *Forest and plantation areas at 2000 in the case study countries (millions of hectares)*

	Australia	Chile	China	India	New Zealand	South Africa	UK	World
Forest plantation	1.4	45.1	2.0	32.6	1.5	1.5	1.9	186.7
Percentage of total forest	0.9	27.6	12.9	50.9	19	16.8	67.9	4.8
Natural forest	153.1	118.4	13.5	31.5	6.4	7.4	0.9	3682.7
Percentage of total forest	99.1	72.4	87.1	49.1	81.0	83.1	32.1	95.2
Total forest	154.5	163.5	15.5	64.1	7.9	8.9	2.8	3869.5

Source: FAO, 2001a

Government motivations

Few countries can claim to have based their policies on the development of plantations over the years on concerted and systematic processes of gauging public opinion. However, it is useful to examine the objectives, stated and derived from practice, of plantation development in terms of their apparent driving forces and sources of demand. Usually, there were multiple motivations that gave state plantation agencies every reason to press on, but caused problems when no priorities were attached or no guidance given on how to balance often competing objectives. India's forestry policy of 1952 listed several 'paramount needs' of the country that were to provide the fundamental basis for forest management. Everything from environmental services to industrial raw material, and from rural subsistence requirements to revenue for the government, was included in the list. Although it was obvious that not all of these objectives could be met simultaneously, there were no guidelines as to how choices were to made between these competing claims on the forests (Vira, 1995).

Ensuring timber supply

Creating a strategic reserve of timber for national security in the event of war preventing imports was the main reason for the UK government establishing the Forestry Commission in 1919 and giving it the task of creating a strategic reserve of timber. That objective remained part of successive governments' policies until the 1950s, when it was dropped after the proliferation of nuclear weapons made it likely that a future war would render such strategic reserves meaningless. Compensating for declining indigenous forest resources has featured in South Africa's, New Zealand's, Chile's and India's forestry policies. South Africa's first plantations were established in the former British-governed Cape Colony as an alternative to the fast disappearing small area of natural forests and expensive imported timber. With a growing economy, the Boer War and a rapidly expanding mining industry there was an increasing demand for construction timber. India's government began to establish plantations during the 1840s to compensate for the removal of valuable timber species from natural forests.

Import substitution

Import substitution was one of the Australian government's main objectives of the early phase of plantation development, and was a driving force for the development of state plantations until the late 1970s; considerable funding was provided to establish a plantation resource for that purpose. In the UK, in addition to the national security argument, import substitution became an objective of government policy until the 1970s. It was knocked on the head when the government recognized that the UK did not have a comparative advantage in wood production; it was cheaper to import timber and invest, instead, in economically more productive activities. The argument has been revived recently under the more respectable guise of reducing pressure on natural forests in other countries.

National economic development

Large-scale timber plantations have traditionally been valued for their potential – though not often realized – contribution to industrial development. This was true of plantations in Europe as early as the 18th century, and more recently in developing countries. Timber was – and continues to be – a critical input to a number of industries from the railway sector, to construction business and paper production. 'Trickle-down' theories of plantations and industrial development became increasingly prominent during the 1950s and 1960s (Sargent and Bass, 1992). In addition to supporting local industries, timber has been seen by many as a key export, providing valuable foreign exchange to support growing international trade. As an example, from an effective base of 0 during the 1940s, South Africa has developed a plantation estate that produces 1.63 per cent of global pulp supply, 0.76 per cent of global paper supply and 0.3 per cent of global sawn timber supply, and contributes US$1 billion annually to the economy (1.5 per cent of GDP). New Zealand's and Chile's plantation estates are even more significant in their national economies, contributing, respectively, 4 and 3 per cent of GDP.

Rural livelihoods

In all of the case study countries, at different times employment has been an explicit objective. During the 1930s, South Africa's plantation programme was aimed at job creation for unemployed whites during the depression years. Later, a 1956 government commission into socio-economic development recommended that forestry be used as a regional economic development instrument in the areas that were to become the racially defined 'homelands'. Implementation of the initiative became the responsibility of the different homeland administrations and 150,000 hectares of plantations were established. The New Zealand government used the New Zealand Forest Service (NZFS) to try to deliver social objectives, particularly reducing rural employment as New Zealand's unemployment figures climbed to politically embarrassing levels (Birchfield and Grant, 1993). In the UK during the 1920s and 1930s, the Forestry Commission created 'forest villages' to house workers recruited to establish plantations in economically depressed rural areas. Various land settlement schemes grew into the Forestry Commission's Forest Workers' Holdings initiative to integrate forestry with agricultural employment. Employment grew in this pre-mechanization phase of plantation establishment.

Using plantation development as a vehicle for job creation has created problems. In the UK by the late 1960s, employment objectives were in increasing conflict with job shedding, which occurred over the government's whole estate in response to mechanization, altered forestry practice and, above all, the pressure to show positive financial returns. From this period, the trend was to move from direct employment by the state to the use of self-employed contractors who tender for work. These contractors often led nomadic lives, typically living in caravans as they chased the work from place to place. Today, the peripatetic nature of this workforce, combined with highly machine-intensive forestry practices and the rarity of local forest ownership, has effectively de-linked forestry from local development (Mayers and Bass, 1999). New Zealand's policy of job creation led to significant overstaffing of the NZFS particularly in many economically depressed rural regions. The NZFS was seen as a 'job for life' for many who joined it and considerable institutional barriers existed to dismissing staff (Birchfield and Grant, 1993). South Africa's homeland governments used their powers to develop plantations as a vehicle for rural job creation, resulting in high levels of overstaffing and a lack of commercial focus, which have been carried through to the present day.

Claims of rural development benefits from plantation development are contested. For example, in New Zealand and Australia some members of rural communities have believed plantations to have had negative impacts upon agricultural enterprises and rural social structures (Aldwell, 1984; Le Heron and Roche, 1985; Roche, 1990b – all cited in Schirmer, 2002a; Schirmer and Roche, 2003, see Chapter 9).

Diversifying the rural economy

Diversification of the rural economy has been a significant driver of government support for plantation development in New Zealand and the UK. During the 1980s, New Zealand's government actively promoted farm forestry development through loan schemes and providing supporting legislation. Since the late 1980s, there have been no direct government incentives encouraging farm forestry; but since privatization of the state's plantation estate, the large part of new plantation establishment has been small-scale plantations established by farmers, Maori and other groups on rural land, effectively making plantations a significant contributor to diversification of rural enterprises. In the UK, plantations were explicitly stated as being necessary alternatives to agriculture during the late 1980s when agricultural surpluses were seen as an embarrassment in the context of EU-wide policy, and continue to be important elements of the rural development policy (Forestry Commission, 1998; 2000).

In Australia, government policy at federal and state levels is generally to ensure a level playing field for different agricultural enterprises, of which plantations are one. While diversification of the rural economy is not an explicit aim, federal and state governments have actively funded a wide range of programmes that encourage development of farm forestry – generally defined as landholders actively establishing plantations on part of their property as a component of a mixed agricultural enterprise. The area of farm forestry is increasing but remains a small part of the total plantation estate, with the majority made up of state and corporate or private business plantations. There are also many joint venture and lease plantations in which a farmer/rural landholder provides part of their land and a private company or government agency provides capital and knowledge and establishes and manages the plantation, paying the landholder an annual rent or proportion of the income from the harvested crop (see Chapter 5).

Energy

Wood was the main source of energy in all of the countries studied until fossil fuels were discovered and the technology was developed to exploit them. Wood continues to be important in developing countries where supply of alternative sources is insufficient or inaccessible. In China, forests are an essential source of energy for 40 per cent of the rural population. Fuelwood is by far the most important product extracted from India's forests. Of the total demand for wood in the country, it is estimated that over 80 per cent of the demand is for fuelwood, with a substantial proportion met by unsustainable removals from forests and plantations (Saigal et al, 2002).

As early as the 1970s, India's National Commission on Agriculture (NCA) suggested that plantations should be established on non-forest lands in order to reduce local communities' dependence on state forests, including for their fuelwood needs. It was also hoped that the increased supply of fuelwood from these plantations would reduce the use of cow dung as fuel so that it could be used as manure in agricultural fields. According to estimates, over 458 million metric tonnes of wet dung were being used annually as fuel. If this was used as manure, it would potentially fertilize 91 million hectares and increase food output by 45 million metric tonnes (Srivastava and Pant, 1979, in Pant, 1979).

In South Africa, the Department of Water Affairs and Forestry (DWAF) is responsible for approximately 100 community woodlots planted for fuelwood and other subsistence products. There are many other such plantations nationwide being managed by provincial agriculture departments or by communities themselves. In the UK, wood for energy is promoted in regional forestry strategies (see Box 2.1).

Environmental services

Governments have promoted and paid for plantation development in order to provide a variety of environmental services. The UK government added recreation to its forestry policy objectives during the 1950s in response to

---

Box 2.1 Welsh National Assembly's policy on energy from wood

The development of a vibrant renewable-energy sector in Wales is a key target for the National Assembly, as part of our responsibility to respond to global climate change. Wood has a number of advantages as a fuel, particularly where such wood fuel is an additional product from the management of woodland. We believe that the development of wood-fuel technologies can make an important contribution to the production of renewable energy in Wales. The development of facilities to burn wood at appropriate locations where there are constant demands for heat, such as community facilities, hospitals or industry, can provide both power and background heating. These heat and power plants will contribute to our target of maximizing energy efficiency in Wales:

*   We will integrate energy from wood fuel into the assembly's renewable-energy strategy, ensuring that it is recognized as a key fuel.
*   We will work with ... partners to develop information and advice for Local Planning Authorities on wood-based renewable energy.

Source: Forestry Commission, 2000b

increasing demand for countryside access fuelled by increasing leisure time. Enhancement of rural landscapes followed soon after, and from the 1980s onwards derelict post-industrial and urban landscapes were enhanced, with a 'boom' in the development of 'community forests' to provide attractive and healthy environments for urban and urban fringe communities. Expansion of the UK's planted forest estate is also promoted as enhancing biodiversity; a number of previously rare bird and mammal species have prospered in the UK's new forests. There have also been losses owing to planting of valuable open ground habitat and conversion of semi-natural forest; current policy and instrument directs plantation establishment away from these habitats.

Plantations are promoted for land reclamation purposes – for example, stabilization of dune mining operations in South Africa with *Casuarina*, and woodland establishment on the sites of former gravel workings and coal mines in the UK.

Watershed protection is an important element of China's forestry policy. Following the establishment by the central government of the Forest Environmental Benefit Compensation Fund to pay for forest restoration, local governments have demonstrated their willingness to raise funds for improved land management and forestry to combat soil erosion and reservation siltation by charging water consumers (Lu Wenming et al, 2002). There are contra-arguments, however. In South Africa, plantation forestry is a taxed 'stream-flow reduction activity'. In the UK, large-scale afforestation of sensitive catchments has been associated with acidification of lakes, causing reductions in fish populations.

Plantation development for combating land degradation has been promoted by the government in New Zealand, where it is generally accepted as having been a success (MAF, 2002b), and in India where more than half of the country's area (175 million hectares) is subjected to different types of degradation (GoI, 1999).

Governments are creating a favourable policy and market environment for the development of plantations for combating climate change through carbon sequestration, though not all governments are in favour; China has been amongst those opposing the inclusion of forest-based carbon offsets in the Kyoto Protocol (Lu Wenming et al, 2002). Plantation development has the potential to offset only a small proportion of greenhouse gas emissions, and its greatest potential probably lies in providing a renewable energy source (DETR, 2000). The UK government introduced the Non-fossil Fuels Obligation, which requires electricity supply companies to purchase a proportion of their energy demand from renewable energy sources and enables electricity generators to bid to supply the renewable quota. The development of electricity from wood has not been without its problems. The much vaunted Project Arbre at Eggborough in the UK, an 8 megawatt (MW)-generating plant designed to consume short-rotation willow coppice, went into liquidation after it failed to demonstrate its profitability, leaving 40 staff redundant and 50 farmers with no market for willow crops grown under a 16-year contract with the company and grant aided by the government.

Private-sector motivations

Private-sector motivations have been encouraged by government incentives and supportive tax policies in many countries; but these have not been a precondition for private-sector plantation development everywhere. Traditional private estates in the UK established plantations for game, timber and private amenity long before the government gave any support for planting. Mine owners in Chile began establishing plantations to supply pit props during the early 1900s. More recently, as large multinational players have entered the arena, investment in plantations for profit from wood sales or to provide timber for vertically integrated businesses has emerged without government support. The trend is most apparent in countries with comparative advantage in growing timber owing to high growth rates and low land and labour costs. In some, private-sector motivation is about 'staying in the game' as government or market pressure requires companies to demonstrate responsible stewardship by certifying their performance. Some companies are recognizing new opportunities in emerging markets for environmental services, such as carbon sequestration.

Direct payments by government

All of the case study countries have offered payments in the past for plantation development and management. Over time there has been a clear shift in the types of goods and services that governments are seeking to purchase, with far greater emphasis now on social and environmental services. South Africa has stopped offering direct payments, with Forestry SA (South Africa's national industry organization) expressing the view that the private sector is not looking for direct incentives from government. Where these existed in the past, they created distortions in the market, reduced efficiency and a lack of competitiveness, which was not good for the long-term interests of the industry (see Chapter 10).

In the UK, the grants offered to private land owners by the Forestry Commission were traditionally aimed at purchasing hectares of plantation and cubic metres of timber. This focus continued for 60 years until opposition to large-scale planting of non-indigenous species become so strong that the commission introduced requirements into its grant schemes that effectively meant it was paying for environmental and social benefits instead of timber. Planting grants were tailored to promote a much greater variety of plantation types; there was more emphasis on broadleaves, small-scale woods, community woods offering public access and recreation of woods, and forests of native species. Correspondingly, there was less on large-scale afforestation of unimproved land with non-native species. The concept of 'public money for public benefit' caught hold (see Chapter 11).

Plantation developers in Chile have access to planting and management *subsidies* if the land is classified as preferentially suitable for forestry and the owner is classed as a small property owner. Subsidies for large ownerships were removed in 1974. Owners also have generous tax concessions (see later in this section).

Australia has used grant-loan mixes from state governments and state agencies to encourage private planting in the past; but these are generally not offered currently. One example was the Farm Forestry Agreement Scheme offered by the Victorian government from 1967, which offered low-interest loans with repayments deferred for the first 13 years for establishment of softwood plantings. The scheme had fairly low uptake – after 15 years, approximately 8300 hectares had been established – and the scheme had high administration costs. There were also difficulties with repayments when some scheme participants were unable to sell thinnings from their plantations (Hurley, 1986).

In New Zealand, the only government grants offered today for plantations, via the East Coast Forestry Project, aim to encourage planting on erosion-prone land. The grants provided are seen as a way of funding the environmental services provided by the plantation.

Tax incentives

Although South Africa provides no direct government payments for plantation development and management, forestry is subject to general income tax provisions, which allow companies to write off income from other activities against their forestry (or any agricultural) interests. Chile offers generous tax breaks to plantation owners (see Chapter 6). Although profits from the logging of plantations are subject to first category income tax, plantations and the land where they are planted (provided it is classified as being preferentially suitable to forestry) are not computed for inheritance, assignations and donations tax. In the UK all forestry activities were removed from income tax in 1988

Going too far with tax breaks The Australian and UK experience with taxation incentives suggests that they can be a very powerful means of generating significant private-sector investment in some forms of plantation forestry. It is also arguable – and a widely shared view in some plantation regions – that the rapid expansion of plantations which they generated was insufficiently tempered by appropriate land-use planning and conflict resolution frameworks (see Chapter 5). In the UK, opposition to the rapid expansion of plantations by private investors led to government removing the tax breaks that had driven investment for 30 years. The anomaly under which forestry operations qualified for tax relief while income and capital appreciation were almost tax free was removed by taking forestry out of the income tax system. To compensate for the large reduction this produced in the incentive to plant, the planting grants were increased substantially.

Favourable trade and investment environments
None of the countries studied offered trade incentives targeted at the forest sector. Liberal trade policies have helped to stimulate a massive growth in exports in Chile, with consequent improvements to the country's balance of trade and economy. Deregulation of international trade helped to make New Zealand internationally competitive. Prior to the structural adjustment programme implemented by the Labour and subsequent National governments from 1984, a large number of export incentives and import controls were in place (Roche, 1990b; Birchfield and Grant, 1993). Since 1984 most of them have been removed. Various international trade liberalization processes aim to reduce international barriers to trade, including the General Agreement on Tariffs and Trade (GATT), the North American Free Trade Agreement (NAFTA) and the World Trade Organization (WTO) (Walker et al, 2000; Turner et al, 2001). The impact has been to open New Zealand wood processors and log sellers to international markets. Deregulation (through deregulation of the ports and removal of trade barriers such as import tariffs) was successful because it was accompanied by a range of domestic reform measures that enabled a more internationally competitive forest sector to develop.

Australia's eucalyptus plantations are often funded by international investors attracted by the low sovereign risk and good growing conditions, although many are not committed for sale to particular international customers/ investors. There is ongoing debate about the optimal level of value adding that can/should occur for export.

New Zealand's government created a favourable investment environment by establishing a *Forestry Right* to enable joint ventures to take place between investors/capital providers and landholders. The Forestry Rights Registration Act 1993 provides for a *profit à prendre* Forestry Right, which is a right granted to the landholder, other person or entity to establish, maintain or harvest a crop of trees along with ancillary rights of access and works needed to undertake plantation activities (Forestry Joint Venture Working Group, 1991).

Markets for forest environmental services
Market and payment mechanisms have emerged, or are showing signs of emerging, around some plantations. Landell-Mills and Porras (2002) reviewed 287 case studies of emerging markets for forest environmental services. The following sections draw from their study.

Markets for carbon sequestration Payments for carbon sequestration can help to finance continued provision of public forest goods and services (see Box 2.2). When carbon credits are tied to forest management standards, they can help to enforce continued provision. Markets are developing rapidly, though evidence that small growers in developing countries face serious constraints in

---

Box 2.2 Development of the carbon market in Australia

There have been a number of agreements in which carbon rights have been registered over areas of plantations. For example, State Forests New South Wales (SFNSW) has an agreement with Tokyo Electric Power Company (TEPCO) for the establishment of plantations – up to 40,000ha over ten years – for which TEPCO will own both the trees and the carbon rights (SFNSW, 2002). Similarly, Western Australia's forest protection committee (FPC) has an agreement with British Petroleum in which plantations are established to offset carbon emissions (FPCom, 2001). While many in the forests' sector believe that there is considerable potential for future markets to develop in which carbon rights for plantations are traded, this has yet to occur – despite some pioneering attempts by the Sydney Futures Exchange (SFNSW, 1999).

Several states are preparing for the emergence of environmental services markets by enacting relevant legislation. For example, in May 2001, the Victorian government enacted carbon property rights legislation, which also supports investment in environmental plantings for purposes such as habitat expansion, mitigating salinity and land protection (NRE, 2002). Some government agencies have also sought to develop investment structures that would capture these environmental services markets in support of tree-growing, particularly in lower rainfall zones (e.g. Salvin, 2001).

Source: adapted from Schirmer and Kanowski, 2002

accessing market opportunities is cause for concern. The potential use of the forestry-based offset market is still very dependent upon policy decisions and upon how they will be accounted for (FAO, 2001g).

Markets for watershed protection Watershed services benefit groups of individuals and are characterized by threshold effects. Cooperation in demand and supply is therefore key. Market development depends upon strengthening

Box 2.3 Payments for watershed protection in China

Unilever's Clean Water and Green Mountains for China Initiative

Unilever – a major multinational producer of consumer durables – estimates that worldwide it consumes 0.1 per cent of total water extracts for use each year. In recognition of its major role as a water user it runs a Water Care programme that aims to 'ensure that [Unilever's] activities and those of [their] suppliers, customers and consumers achieve a sustainable balance ... so assuring the ability of future generations to access sufficient quantities of clean water.'

As part of its programme, Unilever manages a number of local level initiatives. In June 2000 it launched its Clean Water and Green Mountains for China Initiative. The initiative represents a long-term commitment by the company to pay US$845,000 a year towards reforestation and soil and water conservation efforts in water-stressed areas of China. During its first year, Unilever planted 500,000 trees. In 2001, tree planting was expected to reach 1 million trees.

Xingguo County, Jiangxi Province

In 1980 soil erosion affected 85 per cent of Xingguo County, an area of 190,000ha. As 96 per cent of the forest lands in the area have been contracted out under the Household Responsibility System, households have been given support to plant and manage trees for soil conservation. Over 50 per cent of the amount, to date, has come from household and private sources. Private industry is forced to contribute through fees. The metallurgic industry pays 0.5 per cent of sales revenue; the chemical industry pays 3 per cent of sales; coal enterprises pay US$0.01 per tonne of output; and hydropower companies pay US$0.0001 per kilowatt output. The result of this investment is dramatic. By 1999 the area affected by serious soil erosion had dropped by almost 80 per cent, to 41,000ha.

Source: adapted from Lu Wenming et al, 2002

cooperative and hierarchical arrangements to allow beneficiaries to formulate group payment strategies and to tackle the difficulties of excluding non-payers from watershed services.

It is not clear whether the market provides a preferable mechanism to tried and tested regulatory systems. The lack of attention to equity impacts of emerging payment schemes raises a number of concerns.

Markets for landscape beauty The market for landscape beauty remains relatively immature. Constraints to market development are well established and shifts in power balances are difficult to make. As long as tour operators resist paying, land stewards' opportunities for being rewarded for the services they provide lie in establishing themselves as marketing enterprises. To do this they need skills to administer and manage complex international businesses. In the plantation sector, markets have been effective in purchasing tree-planting as part of major industrial or residential developments or urban improvement initiatives in joint ventures between private companies and not-for-profit forest management organizations such, as the Woodland Trust and community forest initiatives in the UK. They are unlikely to emerge as drivers of large-scale plantation development.

Markets for biodiversity conservation It is not easy to commercialize the diversity of nature. Services provided by biodiversity are numerous and most are intangible. Services are rarely consumed by a clearly identifiable clientele and it is difficult to apportion the services to different buyers. In spite of these problems governments, international NGOs and private companies are paying for biodiversity conservation, driven by growing public awareness of biodiversity benefits and threats of loss. Individual and community land stewards have become increasingly proactive sellers of their services.

But the markets for biodiversity services remain nascent. There are significant transaction costs associated with setting up and implementing trades. There are also implications for distribution of costs and benefits. Early indications are that the livelihoods of poor communities may be threatened by the market through increased exclusion, lower incomes and a weaker asset base.

Governance of sustainable plantation development and management

Plantations have many benefits, but they also have potential negative social and environmental impacts (some of which we have already highlighted) and which governments will want to avoid or mitigate, though during the past they have often ignored or been slow to react to problems caused by their own policies and actions. Social impacts include:

- displacement of rural people;
- reduction or elimination of access to products and services;
- changed power structures in communities.

Environmental impacts include:

- reduction in stream flow and water yield from catchments;
- soil erosion due to poor road layouts and drainage lines;
- biodiversity loss where plantations are created on valuable open-ground habitats or following conversion of natural woodland;
- reduction in landscape quality as a result of poor plantation design.

Examples of the social and environment impacts that have occurred or have been anticipated in the case study countries are presented in Boxes 2.4 and 2.5, respectively.

Mayers et al (2002) observe that the attainment of sustainable forest management depends critically upon the extent and quality of enabling policy, and legal and institutional conditions. Together, these conditions influence how a society organizes itself to develop and manage forest wealth, to produce forest goods and services and to consume them. Weak forestry institutions cannot enforce legislation. Weakened social norms mean that forest abuse is unpunished by other stakeholders. These observations apply as much to attaining sustainable plantation development as they do to attaining sustainable management of natural forests. Building blocks of an effective forest governance framework include public policy, the system of tenure and use rights, the design and functioning of forestry-related institutions, and the mix of legislation and other instruments for delivering policy goals. We look at each of these, in turn, in the next sections.

Sending the right signals through clear public-policy goals

Clear signals are important for private actors and government promoters and regulators. The case study countries have taken different approaches. Some have policies on plantation development that explicitly address the negative impacts while promoting the benefits of plantations. Those that have decided to leave plantation development to the market have nevertheless addressed potential negative impacts through regulatory instruments. The following examples illustrate the variety of approaches.

Australia. In general, tree-planting for commercial plantations and tree-planting for environmental rehabilitation have been separated in government policy, though this is now changing in response to the imperative of commercially viable land-use systems to address salinity (e.g. oil mallee eucalypts in Western Australia). Commercial plantation policy is addressed primarily by the *2020 Vision for Plantation Forestry*, launched in 1997 in which the federal and state governments and forest industries set a goal

Box 2.4 Social impacts of plantations: Displacement and exclusion of black communities in South Africa

The key forestry policy objective developed by the South African government, particularly from the 1930s onwards, was one of plantation expansion. Generally, sites were selected according to conventional silvicultural considerations. However, much of the establishment took place in the wider political context of 'separate development' for blacks and whites, which resulted in the formation of the so-called 'homelands'. Forestry was one of the many land uses used by the apartheid government as a context for relocating large numbers of people from their traditional areas of occupation to the newly formed homelands. The instruments available to the government of the day for expanding plantations were racially discriminatory. Applicable land legislation would have enabled the 'reservation' of communally owned land by the state for forestry purposes. This resulted in widespread dispossession of land. The form of governance under the apartheid regime was centralist and top down. This same approach was adopted within the homelands by the various puppet governments that were installed by Pretoria. There was no culture and history of consultation. Decisions taken centrally were enforced locally. For this reason, there is a significant body of resentment against plantations in certain areas to this day.

Once the plantations were established, the form of governance within the forest service continued to be exclusionary and top down. Local people did not participate in management and were not consulted for the purposes of decision-making. They may (formally and informally) have enjoyed some access and use rights for forest products, such as fuelwood and building materials, and some compartments were set aside to produce poles for the community; but this was at the discretion of forest managers and was unevenly applied across the country. The culture within government, down to the level of forest guard, was hierarchical and, to some extent, remains so. This culture is not unique to South Africa and its apartheid history, but is commonly found in post-colonial forestry services worldwide.

There was no participation by black people in the economic development of the plantations and the processing industries which grew up around them. All the so-called 'evergreen' long-term timber supply contracts were with white-owned companies, who more or less enjoyed exclusive access to a cheap resource produced on communal land for which the community was not compensated.

Black people were employed in the plantations, but usually under white managers. To this day, the government has a huge excess of low-skilled black staff and a shortage of black managers and supervisors. This pattern reflects a lack of human resource development for black staff within the forest service. Despite the best of intentions, this mismatch of skills has not been successfully addressed since the end of apartheid in 1994.

Several small (mostly coloured-owned) 'bush-millers' (mobile sawmillers) emerged to harvest and utilize the timber which the large contract holders could not use. These small millers were, however, never given any security of supply and rarely developed into anything more than the survivalist operations that we see today.

Source: Dlomo and Pitcher, 2002

Box 2.5 Environmental impacts of plantations: Experience from Australia, India and New Zealand

Australia. Establishment of largely mono-cultural stands of exotic species is seen as having some positive and some negative environmental impacts, depending upon the locality and management of the plantation. Positive impacts may include reducing the risk of dryland salinity, reducing soil erosion or otherwise rehabilitating land. Negative impacts include concerns about biodiversity loss (especially where plantation establishment occurs on natural or semi-natural habitat), levels of water use and impacts on landscape (Schirmer and Kanowski, 2002).

India. The social forestry programme of the 1980s resulted in the conversion of large areas of natural forest to plantation. Large-scale plantation of Chir pine (*Pinus roxburgii*) in the Himalayas has been blamed for depletion of soils and streams, suppression of undergrowth, destruction of indigenous oak ecosystems, and displacing local biodiversity upon which local people depended for their livelihoods (Sinha, 2002, in Danodaran, undated).

New Zealand. Concern about sustainability of indigenous forest logging and conversion of indigenous forest plantations was expressed by a growing environmental movement during the country's second planting boom. The types of concern were exemplified by the opposition to the Beech Scheme, in which a proposal to convert some beech forests to exotic plantations as part of a broader scheme of logging indigenous beech forests was strongly protested, with a 340,000-signature petition opposing the scheme presented to Parliament in 1975. The scheme was eventually rejected (Schirmer and Roche, 2003).

of trebling Australia's 1997 plantation estate by the year 2020. The *Vision* includes a commitment to removing impediments to developing plantations and providing a conducive investment environment. Commercial plantations have not figured prominently in sustainable development strategies, which focus primarily on non-commercial re-vegetation strategies, but which are, in general, promoted by federal and state governments as a useful example of developing sustainable and profitable rural industries.

Chile's forest plantations policy has a strong economic focus. It is aimed at the formation of a sizeable forest resource base to enable the expansion of an export-orientated industry and to contribute to the country's economic growth. The development of the sector is nevertheless controlled by various regulations designed to avoid or mitigate negative impacts.

India. Key elements of India's National Forest Policy (GoI, 1988) are presented in Box 2.6. The policy places great emphasis on increasing tree cover in the country. It states that the national goal should be to have a minimum of one third of the total land area of the country under forest or tree cover. However, it is envisaged that most of the new plantations will be created in degraded forest and non-forest lands and there is a presumption against making natural forests available to industries for creating plantations. The National Forestry Action Programme (NFAP) mentions the need for a plantation master plan for the country. This plan would cover the following

---

Box 2.6   Clear signals in India's forest policy (at least from one key part)

India pursued a policy of creating large-scale industrial plantations on state forest lands and social forestry plantations on non-forest land from 1976. The policy failed to deliver and met with stiff opposition. A new forest policy was issued in 1988, which signalled a radical new direction. It stated that conservation and local communities' needs should be the major objectives of forest management and that industrial plantations should not be encouraged on state forest lands:

> The principle aim of Forest Policy must be to ensure environmental stability and maintenance of ecological balance, including atmospheric equilibrium, which are vital for sustenance of all life forms, human, animal and plant. The derivation of direct economic benefit must be subordinated to this principle aim (GoI, 1988).

> The life of tribals and other poor living within and near forests revolves around forests. The rights and concessions enjoyed by them should be fully protected. Their domestic requirements of fuelwood, fodder, minor forest produce and construction timber should be the first charge on forest produce (GoI, 1988).

> As far as possible, a forest-based industry should raise the raw material needed for meeting its own requirements, preferably by establishment of direct relationship between the factory and the individuals who can grow the raw material ... the practice of supply of forest produce to industry at concessional prices should cease. Industry should be encouraged to use alternative raw materials. Import of wood and wood products should be liberalized (GoI, 1988).

> Natural forests serve as a gene pool resource and help to maintain ecological balance. Such forests will not, therefore, be made available to industries for undertaking plantation and for any other activities (GoI, 1988).

> No such [plantation] programme, however, should entail clear-felling of adequately stocked natural forests. Nor should exotic species be introduced, through public or private sources, unless long-term scientific trials undertaken by specialists in ecology, forestry and agriculture have established that they are suitable and have no adverse impact on native vegetation and environment (GoI, 1988).

Source: adapted from Saigal, 2002

areas: an inventory of the existing plantations; general prescriptions and guidelines for managing existing plantations and rehabilitation, wherever required; and preferred locations for new plantations and the species to be used.

South Africa. The drive for inclusion and equality that followed the 1994 elections required far-reaching changes in the policy and legislative framework for forestry (see Box 2.7). The White Paper of 1996 (DWAF, 1996) and

---

Box 2.7 South Africa's enabling policy and legislative framework

Significant changes have taken place in terms of the policy and legislative framework for forestry since the end of apartheid. Democratic principles are built into government policy, along with an explicit policy intent to address the injustices of the past. A new Forest Act and White Paper provide an enabling environment for the sustainable management and development of forests and for the government forestry service to engage with all stakeholders in a participatory and empowering way. The act created a framework for privatization of state forests. Specifically, the Forest Act provides for public access to all state forests for cultural, recreational and spiritual purposes and exempts local communities from requiring a licence to harvest forest products for domestic use. These provisions remain in place even after privatization.

It takes time, however, for the culture of an organization to catch up with its own policies. There are still many examples of management approaches on the ground remaining much as they were 30 years ago. However, participatory approaches are filtering through the government forest service. This has resulted in the Directorate of Participatory Forest Management, which is set to take forward all aspects of community participation in the forest sector and which will work closely with the office responsible for restructuring and privatization. The privatization process is one in which the government has attempted to put its democratic principles into practice. Although there is a long way to go, and many detractors believe that more empowerment could be achieved, the process has been heralded widely as a successful example of how government can act in a transparent, participatory and empowering manner.

The act also requires the minister to develop criteria, indicators and standards for sustainable management of all forests. There is an ongoing consultative process to develop the criteria, indicators and standards for different forest types. Those that will apply to commercial forestry plantations are close to finalization. The process has received widespread support from the private sector, which is not altogether surprising when South Africa boasts that 85 per cent of its plantation area is already certified, the highest of any country in the world.

Source: Dlomo and Pitcher, 2002

the Forestry Act of 1998 (DWAF, 1998) contain clear statements of intent with regard to, among other things, access to and ownership of forest land and engagement with stakeholders by government, and provide the powers necessary to make change happen.

United Kingdom. The UK's forestry policy has changed in scope and emphasis many times since embarking on its programme of reforestation in 1919. The clarity of the objective of creating a strategic reserve of timber with the powers and money to deliver enabled the Forestry Commission to quickly develop a sizeable plantation estate. The broadening of the commission's mandate to include recreation, landscape and later the entire spectrum of forestry benefits presented as multipurpose forestry was more of a strategy by the commission to stay in business than a considered policy of the government to address people's needs and aspirations. As the commission became 'all things to all people', it became more difficult for its staff to balance its many competing objectives, to establish a clear vision and to determine how well the organization was performing. Devolution of forestry policy to the new Scottish and Welsh governments in 1997, coupled with devolved control over the commission's delivery of policy, was an opportunity for fresh thinking and clearer direction towards locally important objectives. The vision developed by the Scottish government is set out in Box 2.8.

Tenure and use rights

Tenure and use rights determine access to, and distribution of, the benefits that flow from plantations. Governments can play a decisive role in ensuring continuing access to traditional rights and equitable distribution of benefits. Mayers et al (2002) propose essential ingredients for good governance in relation to tenure and use rights.

Rights must be clear and defensible. Land and property tenure needs to be secure, clear, documented and non-discriminatory against forestry. There need to be clear, equitable and legally defensible rights in place: rights to manage the forest resource (based on free and informed consent of others with legal and customary rights); and rights to extract resources from public forests given in return for full economic compensation, including externalities. In addition, stakeholders need to be aware of their rights and the avenues open to them to contest them (Lindsay, 2000).

Rights must be exclusive. Holders need to be able to exclude or control the access of outsiders to the resource over which they have rights. There must be certainty about the boundaries of the resource to which the rights apply and about who is entitled to claim membership in the group (Lindsay, 2000).

New law may be needed to back up rights. Examples from India and elsewhere testify to frequent confusion about the way in which benefits are to be shared, leading to false expectations and possible disillusionment. Vietnam has made new laws to support a policy of allocating forest land to rural households and allowing these households to share in benefits from forest management and protection. The laws provide for the long-term allocation of forest land

Box 2.8 Scotland's vision and guiding principles for forest development and management

The Scottish Forestry Strategy is the Scottish Executive's framework for taking forestry forward, through the first part of the new century and beyond. The vision is that Scotland will be renowned as a land of fine trees, woods and forests which strengthen the economy, which enrich the natural environment and which people enjoy and value. High-quality trees, woods and forests can help to make Scotland a better place for people to live and work in and to visit. Increasing confidence in the future of forestry will encourage investment that will benefit current and future generations. Scottish forestry can do this by:

- promoting sustainable economic growth, taking advantage of Scotland's competitive advantages in terms of resources and infrastructure;
- enhancing the environment and heritage, enriching and extending important woodland habitats and creating attractive landscapes in both town and country;
- helping to create a more inclusive society, offering opportunities for employment – particularly in rural areas – and for enjoyment, as well as providing 'lungs' for cities and towns.

A number of *principles* form the basis for the development of this strategy and its implementation. The overarching principle for the strategy is sustainability. Scottish forestry must contribute positively to sustainable development and meet internationally recognized standards of sustainable forest management.

The next principle is integration. Scotland's woods and forests do not exist in a vacuum. Forestry should fit well with other rural activities in Scotland, such as agriculture, conservation, deer management, fishing, recreation and tourism. The overall objective should be sustainable rural development, rather than the promotion of individual activities in isolation from one another. Integration needs to build on the success of current partnership projects, working together to tackle complex and difficult issues.

Forests and woodlands should contribute to the well-being of the people of Scotland. This is the principle that they should contribute positive value. This contribution may be in social terms, in economic terms or in environmental terms – and there should be benefits that clearly exceed costs. Although they can be difficult to measure, it is important that the value of non-market outputs is recognized.

Forestry should earn community support. Forests should be managed in ways that enjoy broad public support. 'Community' here includes the local people most directly affected, communities of interest (for example, ornithologists and archaeologists) and the wider Scottish community. Complete agreement may not always be possible; but there should be mechanisms for participation, for sharing and explaining views, and for working towards consensus.

Forestry should reflect the rich diversity and local distinctiveness of the land, forests and people of Scotland. It is important to protect, manage and enhance the rich and varied range of woodland habitats and species, recognizing that different types of forest will provide different benefits and suit different places.

Source: Forestry Commission, 2000a

to households and individuals for silvicultural production, including natural and planted forests and nurseries. Land-use rights allocated to households or individuals may be exchanged, transferred, leased, inherited or mortgaged. The laws specify the criteria of forest land allocation and forest users' rights and obligations, and assign administrative responsibilities and define offences and penalties (Phoung, 2000).

Strong, smart, joined-up government – and other key institutional roles

Effectiveness of policy development and delivery is determined by a host of connections within and between institutions. Poor connections lead to bad policy, conflicting messages, fuzzy accountability, weak enforcement of regulations and inconsistent decisions in relation to licensing and grant aid.

Coordination among government agencies. While government roles may vary, it is critical that interventions are coordinated. When the mandates of different parts of government overlap or are not 'joined up', it will be difficult to ensure the desired balance between market forces and public requirements. Under the Australian federal system, poor coordination between the three levels of government – national, state and local – is particularly problematic (see Chapter 5).

Separating government's management and regulatory roles. Policy development and regulation and management of production processes and service delivery do not sit well together; they require completely different approaches and skills and often come into conflict. Separation of government functions and business functions is a well-established trend in all sectors where governments have played a business role. In some cases, the result is privatization (e.g. electricity generation and supply, water supply and telecommunications). Where governments see a continuing benefit from state management of an asset (public highways) or service (health), they establish management entities separated from the central government organs that have policy and regulatory responsibility for the sector. The same trend is evident in the forestry sector. In the UK, this was part of the 'settlement' when the government rejected wholesale privatization (see Chapter 11).

Complementarity of central and local government powers. Where central and local government share powers over forest development and management, conflicting objectives and instruments may emerge unless their respective roles are clearly defined and accepted. Problems are likely to be most acute in large countries where it is difficult to achieve effective oversight of forest utilization. China, for example, operates a complex and extensive hierarchy from state level to township through the intermediate administrative levels of province, prefecture and county. The State Forestry Administration supervises the implementation of central government laws and policies, but does not have the authority to enforce these laws and policies directly; its regulatory and monitoring roles have to be implemented through lower-level authorities, who have had to intensify their supervision of management standards as responsibility for forest utilization and management has been decentralized to non-governmental actors. Central control over the national annual allowable cut system, illegal logging and preventing the conversion of forests to agriculture has sometimes been strained by changes made at a local level, especially with respect to forestry taxation and tenure systems (Lu Wenming et al, 2002).

Civil society roles and joint government–civil society roles. Participatory approaches to policy development and implementation are a help, not a hindrance. By engaging with NGOs, private business and local communities, governments can identify with greater clarity the issues that policy needs to address and the instruments that will best suit their purpose. Conflict between government goals and civil society aspirations arise where government does not listen and respond. Participatory approaches facilitate 'buy-in' and can lever in much needed financial support. The partnership approach adopted by the UK's state forest management enterprise has brought UK£20 million of external funding for investment in social and environmental plantation services over the last ten years (see Box 2.9).

Partnerships between companies and communities as mechanisms for governance. Company–community deals offer an additional pathway to effective forest governance. Depending upon the terms of the deal, partnerships can be good for forests, as well as for community livelihoods and for the profitability and standing of the company. (Mayers and Vermeulen, 2002a).

Laws and regulations: Building blocks of effective governance

Developing regulatory regimes and incentive structures to ensure that plantations have minimum negative impact, while not unduly impeding investment, seems necessary. All of the case study countries regulate private plantation development and management, though to varying degrees. Key regulatory themes identified in the seven case studies reviewed for this book are set out in Table 2.3. Specific examples of regulations are set out in the next paragraphs.

Zoning land use. New Zealand's Resource Management Act 1991 gives powers to regional and district councils to prepare, respectively, regional and district plans. Regional plans prescribe rules applying to activities that have effects on soil, air and water. District plans identify key resource management issues and set objectives and targets for the district. District councils have powers to control the use of land for particular purposes. Land uses may be classed as permitted, controlled, discretionary, non-complying, or prohibited in particular zones. Where forestry is classed as discretionary or non-complying, plantation managers have to apply for a resource consent before they may carry out plantation activities. Where forestry is classed as controlled, the district council may impose conditions.

Approval of management plan required before activities are carried out. The UK requires applications for felling licences to be accompanied by a management plan. The plan may be simple or complex depending upon the

Box 2.9 Levering in funding through partnerships: Examples from the UK's Forest Enterprise

Regeneration of former coal-mining areas

In 1992, 12 of Nottinghamshire's coal mines were closed with the loss of over 36,000 jobs. During that time, the Forest Enterprise was trying to expand its land holding around urban areas; but high land prices and a small budget prevented this. The Forest Enterprise forged a partnership with the local authority to restore seven of the closed mines' spoil heaps to 760ha of community woodland and levered in funding from the European Union's structural funds and British Coal. The total cost of the development was UK£1.3 million. The Forest Enterprise's contribution was less than UK£0.1 million.

Community forests

Learning from its success with the Nottinghamshire partnership, the Forest Enterprise set out to develop a partnership and funding package to allow it to play an enhanced role in the national Community Forests Initiative – a programme of 12 community forests aimed at improving urban and peri-urban environments, and outdoor recreation and learning opportunities. By 2002, the Forest Enterprise, in conjunction with three community forests (Thames Chase, Red Rose and the Mersey) had levered in UK£11.4 million, with UK£3.2 million of its own money in a partnership with the Treasury's Capital Modernization Fund, regional development agencies, local authorities and non-governmental organizations.

Source: Garforth and Dudley, 2003

size and sensitivity of the forest and the potential impact of the proposed felling.

Mandatory environmental assessment of plans or operations. The UK is required by the European Union (EU) Environmental Assessment Directive to operate a consent procedure for proposed plantations that are likely to have a significant environmental impact. Applications for consent must be supported by an environmental assessment.

Licensing of plantation operations. The creation of new plantations in parts of South Africa that are subject to water deficits is subject to a licensing system. The UK requires forest owners to apply for a licence to fell trees, except for small volumes.

Safeguarding customary rights. South Africa's 1998 Forest Law grants the public access to all state forests for cultural, recreational and spiritual purposes and exempts local communities from requiring a licence to harvest

Table 2.3 *Regulation in the case study countries*

Regulatory provision	Australia	Chile	China	India	New Zealand	South Africa	UK
Mandatory environmental assessment of plans or operations	√[1]						√
Approval of management plan required before activities carried out	√[2]	√	√				
Licensing of plantation operations	√[3]		√		√		√
Penalties for damaging environmental impacts	√	√		√	√	√	√
Minimum health and safety safeguards for forest workforce and forest users	√	√		√	√	√	√
Minimum employment conditions	√	√		√	√	√	√
Equal opportunities to employment	√					√	√
Safeguarding valuable wildlife, cultural, heritage and landscape areas and features	√	√		√	√	√	√
Maintaining customary rights	√				√		√
Establish rights of access in perpetuity	√[4]				√[5]		√
Establish rights of use in perpetuity	√[6]				√[7]	√	

Notes:
1 Varies by state and size and location of plantation.
2 Varies between states and by activity; different activities have different requirements for approval.
3 Licences may be required in some states; NSW regulates harvesting through environmental protection licences.
4 Varies by state.
5 Only for the duration of the Crown Forestry Licence.
6 Varies by state.
7 Only for the duration of the Crown Forestry Licence.

forest products for domestic use. These provisions remain in place even after privatization.

Tying government support to sustainable management practices

Government financial assistance programmes for plantation development, whether through direct grants or favourable tax regime, provide pathways for avoiding or mitigating damaging social and environmental impacts.

Chile's subsidies come with obligations, with severe penalties for non-compliance. Plantation managers must:

* Submit a technical study of the land and its consequent classification proposal.
* Present a management plan for the properties to the Corporación Nacional Forestal (CONAF) within one year of the date of the certificate approving the classification as preferentially suitable for forestry; any felling or logging operation that takes place in lands preferentially suited for forestry may be done only after approval of the management plan by CONAF.
* Restock felled areas or plant an equivalent area elsewhere.

The favourable tax regime that drove most of the UK's private plantation development from the 1950s to the 1980s came with no environmental or social strings attached. Under a 'gentlemen's agreement' between the Forestry Commission and the private-forestry sector, private developers followed an approvals procedure linked to direct grant aid, the value of which was far less than the value of the tax relief on the cost of establishing the plantation. An enormous area was established by afforestation companies on behalf of cash-rich clients seeking to minimize their income tax payments. The system was bitterly criticized by NGOs and many local authorities for driving the establishment of badly located and poorly designed plantations. Criticism grew after two companies established large plantations without going through the approvals procedure. The system was eventually abolished in 1988.

Promoting voluntary agreements: Self-regulation

There are many examples of voluntary codes of practice, though it is not clear how effective they are without some pressure for, and an effective system of, independent verification. In New Zealand, for example, environmental NGOs have negotiated a series of accords and agreements with the forestry industry, some of which affect plantation management (Salmon, 1993). In addition, forestry groups have developed forestry codes of practice. The New Zealand Forest Accord, signed in 1991 between representative industry groups and environmental NGOs, views plantations as a way of producing wood products while protecting native forests. The accord identifies land where it is inappropriate to establish plantations, ensuring, in particular, that plantations do not replace regenerating or mature indigenous forests (Walker et al, 2000). In 1995 the plantation industry and environmental NGOs signed the Principles for Commercial Plantation Forest Management in New Zealand. The agreement complements the accord and aims to promote understanding between the signatory parties with a view to achieving environmental excellence in plantation forest management. The Logging Industry Research Organisation's New Zealand Forest Code of Practice, developed in 1990 and revised in 1993, provides a means of ensuring safe and efficient forest operations that meet the requirements of sound and practical environmental management.

Verifying responsible plantation management: Certification

Over the last decade certification has developed rapidly internationally as a means of independently verifying the quality of forest management. Certification in one form or another (ISO 14001, the Forest Stewardship Council, or FSC, and the Programme for the Endorsement of Forest Certification, or PEFC, among others) has been, or is in the process of being, implemented in all of the case study countries. The growth of certification has been driven by pressure on manufacturers and retailers to ensure that forests that supply them are being managed responsibly. In its application as a market instrument, certification has caused some improvements in plantation management; but these have occurred in plantations run by well-resourced enterprises that were already practicing good management (Bass et al, 2001).

A recent study of certification's impacts in the UK concluded that, overall, certification is having a positive impact on the standard of forest practice (Garforth and Thornber, 2003). Although the evidence pointed more to improvements to documentation than practice, forest enterprises recognized tangible progress on the ground in the following areas:

- management planning and sustainability;
- biodiversity conservation and monitoring;
- contractor management and operations;
- planning of chemical use;
- health and safety;
- silvicultural practices.

Whether market-driven certification will deliver improvements in plantation management more widely remains to be seen.

Governments are recognizing the utility of certification as an instrument for ensuring compliance with standards by plantation managers with whom they have a contractual relationship. Certification can be made a condition of public funding of plantation schemes, or of long-term lease agreements on properties in which the government retains a controlling power.

Changing roles

Governments have been the pioneers of plantation development, driven by a real or perceived lack of private-sector capacity and technical knowledge, a tradition of government responsibility for, and power over, forests, and centrally planned governance paradigms. The cost, length of investment and risk deterred the private sector from participating in a substantial way in most countries. Private-sector planting had to be purchased with public funds in the form of direct grants or tax breaks.

Governments established state enterprises to acquire land for planting through purchase of the freehold or lease, or to enter into partnership agreements with private landholders under profit-sharing schemes and other arrangements. In many cases, these enterprises were also responsible for promoting, grant aiding and regulating plantation development by the private sector, which stored up problems that later resulted in the separation of functions and, in some cases, greater private-sector involvement in state-owned forests.

With large amounts of public money at their disposal and large areas of cheap land with seemingly little value in any other use, some of these enterprises were able to develop very large holdings. In the UK, for example, the Forestry Commission established nearly 1 million hectares between 1920 and 1980.

With risks and costs offset by government subsidies, a thriving private-sector developed alongside state enterprises which bore the costs of developing establishment techniques on often difficult sites. In countries with favourable growing and trading conditions, governments have been able to reduce and, in some cases, eliminate subsidies and the private sector has continued to develop as, for example, in South Africa (see Box 2.10).

Box 2.10 South Africa's global private players

Sappi Forest Products owns and manages 490,000ha of plantations in Southern Africa. In 1998 these plantations supplied 50 per cent of the fibre required to produce 1.8 million tonnes of pulp. Sappi's South African activities are, however, only part of its total operations. A series of major acquisitions during the 1980s and 1990s have made Sappi a world player, currently manufacturing 5 million tonnes of paper and 3 million tonnes of pulp in plants on three continents. 85 per cent of its sales and 70 per cent of its US$6 billion assets are outside of South Africa.

Mondi is jointly owned by Anglo American, De Beers and AMIC, and manufactures pulp, paper and solid wood products. Mondi manages 440,000ha of plantation. Some 40 per cent of its production is exported. Since the late 1980s, Mondi has been acquiring shares in international companies to develop a global presence.

Source: adapted from Mayers et al, 2001

Since the 1980s, when Chile's government divested itself of its plantation assets, there as been a massive shift away from direct state participation in plantation development and increased private-sector involvement in the plantation estates built up by governments over preceding decades, as contractors, lessees of use rights or as the new owners. The transformation has not been uniform. Australia and the UK, where the plantations developed by the state are still largely in state hands, contrast markedly with New Zealand and Chile, where the state's direct commercial involvement in the industry has almost completely stopped. South Africa is rapidly moving to the same position. As private-sector participation in plantation ownership, use and management has increased, governments have taken on a number of new roles. These range from business and investment facilitator and business partner, through to environmental and industry regulator.

There have been many different drivers behind governments seeking to increase private-sector participation, and governments have pursued different models of participation, ranging from contracting out services, through leasing of use rights, to outright sale.

Drivers of change

The drivers behind transfers to private actors in the countries studied are summarized in Table 2.4.

Philosophy of a regulated market economy. There has been a shift in macro-economic policy towards reducing public debt, stronger control over budget deficits, delivering increasing economic efficiency and improved aggregate welfare through the private sector, and reducing state power and widening ownership.

Removing contradictions between government as regulator and as manager. As we observed earlier, governments have seen the advantages of separating regulatory and business functions in all sectors, leading them, in some cases, to remove the business function to the private sector, and in others to separate state agencies. NGOs' criticism of joint roles and their calls for separation of conservation and commercial roles gave added impetus to the breaking-up of the New Zealand Forestry Service.

Addressing inequality and empowering disadvantaged groups. South Africa's current privatization process has, as an explicit objective, affirmative action for the previously disadvantaged. Black companies are favoured over white companies (subject to a series of other criteria).

Multiple motivations. Governments will usually have several reasons for pursuing increased private-sector participation. In New Zealand, corporatization of the New Zealand State Forest Service's commercial functions was initiated by the Labour government, elected in 1984 following a severe economic crisis associated with the flight of foreign capital from the country and liquid reserves falling to a point where the Reserve Bank warned that it might collapse (Birchfield and Grant, 1993). The new government inherited a sluggish domestic economy, rising foreign debt and high unemployment,

Table 2.4 *Drivers in the case study countries*

Regulatory provision	Australia	Chile	China	India	New Zealand	South Africa	UK
Philosophy of regulated market economy	✓	✓	✓		✓	✓	✓
Reduction in public debt	✓				✓	✓	✓
Removing contradictions between government as regulator and as manager					✓		
Empowerment of disadvantaged groups				✓		✓	
Increased profitability/efficiency of the sector	✓	✓	✓		✓	✓	✓
Increased investment in downstream processing	✓	✓	✓		✓	✓	

and focused on addressing the economic problems facing the country. The structural adjustment programmes undertaken by the new and subsequent governments involved a range of measures, including the following (OECD, 1999; Kelsey, 2002):

- deregulating domestic markets and encouraging free trade and investment;
- deregulating the labour market by encouraging individually negotiated employee contracts, rather than union negotiations and collective bargaining;
- cutting income taxes and reducing government spending;
- reducing the size, role and power of the state through corporatization, privatization, devolution and managerialism.

Subsequent privatization of the New Zealand Forestry Corporation (NZFC) was driven by:

- the government's desire to reduce public debt (publicly stated by the government as being the primary motive);
- the ideology of the government at the time;
- concerns about potential political interference with the NZFC;
- concerns about the NZFC's ability to raise capital;
- a perceived need to ensure security of wood supply through processors purchasing plantations;
- the inability of the NZFC and the Treasury to agree the value of the NZFC's assets.

Change models

Models used by the case study countries to increase private-sector involvement in state forest plantations are shown in Table 2.5.

Outsourcing services. Along the spectrum of alternative models for privatization, outsourcing is the least dramatic. Ownership and overall management control are retained by the state, while particular use and management functions are devolved to private contractors, which may be for-profit companies, communities and households or not-for-profit agencies. In return for their services, contractors receive a cash payment to cover their time and expenses, a share of the income or profits generated by the plantation that they have been enlisted to manage, or rights to some of the products of the plantation for their own use or for sale.

Corporatization. This does not fit neatly on a linear spectrum of increased private sector participation. It is, essentially, an administrative measure to get public authorities to act more like private companies. It is also often viewed as a first step towards full ownership privatization. It often involves increased private-sector participation through outsourcing in order to reduce costs.

Transfer of rights to use the land for a limited period. This model involves a greater devolution of power from the state plantation manager to non-state

Table 2.5 *Transfer models in the case study countries*

	Australia	Chile	China	India	New Zealand	South Africa	UK
Outsourcing of service provision to for-profit companies	✓		✓		✓	✓	✓
Outsourcing of service provision to communities and households			✓	✓			
Outsourcing of service provision to not-for-profit agencies						✓	
Corporatization/commercialization of the state's forestry enterprise(s)	✓				✓	✓	✓
Transfer of rights to use the land to individuals, households and private for-profit companies for a limited period of time	✓		✓	✓	✓	✓	✓
Transfer of the state's interest in the land as well as use rights to the land	✓	✓					

entities than contracting out. With outsourcing of harvesting or management activities, the private contractor continues to work for the state plantation manager. Where use rights are transferred, private harvesters work for themselves. In most cases, this extra degree of freedom is curtailed by conditions built into the agreement, the complexity of the conditions being determined by the scope of the beneficiary's rights. Where the rights are limited to, for example, harvesting of timber, a simple licence may control the volume of wood extracted. Where rights extend to full management control for many years, more complex agreements will lay down silvicultural, environmental and social performance criteria.

Transfer of ownership. The 'ultimate' form of privatization, and that most commonly associated with the concept, involves a transfer of the state's leasehold or freehold interest in the land and with it the right in perpetuity to use, sell or transfer that interest to other users. The state gives up all powers to control the activities of the beneficiary except through laws. There are few instances of governments 'privatizing' their plantation estates in this way; most have opted for a form of use rights transfer.

Outsourcing services: Getting the private sector to do the government's work

This model tends to be favoured where public benefits are felt to be too important to risk handing to private operators (e.g. where a plantation is providing an essential biological corridor between remnant patches of natural forest), but where there are opportunities to improve on the effectiveness and efficiency of in-house service delivery by state enterprises.

Australia. Many services have been outsourced (e.g. establishment and management) and are now subject to competitive tender processes. Harvesting in most state plantations has traditionally been done by the private sector through contract. Other aspects of management have been outsourced by many agencies in recent decades. Forestry Tasmania has shifted from conducting 70 per cent of its direct operations using its own resources prior to corporatization, to contracting out around 70 per cent of its direct operations to the private sector. For some agencies, such as Forestry South Australia, contracting these activities to the private sector was standard practice prior to corporatization. The organization currently contracts out about 50 to 70 per cent of planting, site preparation, weed control and silvicultural operations. The Victoria Plantation Corporation also shifted to outsourcing much of the establishment and management operations to contractors, and Hancock Victorian Plantations Pty Ltd has continued this practice since privatization (see Chapter 5).

China. State-owned forest enterprises began to contract out silvicultural activities during the 1990s. Silvicultural wages became tied to the amount of work accomplished (for example, the number of seedlings removed from a nursery or the number of trees planted), rather than the prior system of payment for time involved regardless of output (Zhang, 2001). Other experiments for wage payment are also being tried. A second common arrangement involves a contract under which individuals or small teams perform certain tasks in return for a fixed payment. Most contracts stipulate the required product, such as required seedling survival rates, although there are many variations. In some cases, the state enterprises provide materials and machinery. In others, the contractors possess their own material and machinery – previously purchased from the state enterprises. In all cases, any profits belong to the contractors who are responsible for organizing the labour and materials for site preparation and planting and for tending the site for the first few years. Another arrangement, similar to share-cropping in agriculture, was developed to transfer some state-owned capital equipment to former forest enterprise employees. Individual workers obtain the use rights to the capital and share the income from its use with the forest enterprise. This arrangement was first implemented for vehicles and small sawmills, but has begun to spread to silvicultural and agricultural operations.

India. While current government policy does not encourage corporate private-sector involvement on state forest lands, some companies do operate in state forests and plantations, especially for extraction purposes, by becoming agents of state-owned forest corporations. For example, Ballapur Industries Sewa Unit is currently extracting 25,000 tonnes of bamboo a year from government forests in Makangiri District of Orissa by entering into an agreement with the state-owned Orissa Forest Development Corporation. Several other companies, including J.K. Corp., Orissa and Ballapur Industries are also operating on forest lands in this manner (Saigal et al, 2002).

South Africa. During recent years, there has been a major trend among plantation companies towards outsourcing of forestry operations. More than 15,000 people are employed by contractors. The South African Forestry Corporation (SAFCOL) has been steadily moving towards contracting out all of its routine operations since 1993. The move from direct employment to contracting has been accompanied by some job losses, as well as a decline in wages and working conditions, resulting in an adverse effect on workers in the short term. The trend does, however, offer an important avenue for creation of new black-owned enterprises in rural areas.

United Kingdom. Outsourcing of a wide range of services has enabled the UK's state forest management agency, the Forest Enterprise, to substantially reduce costs and increase business efficiency, reducing pressure on budgets and enabling it to sustain investment in recreation and environmental services.

Limits to what can be contracted out
In the cases reviewed, state plantation agencies generally use their own staff to plan, manage and audit plantation operations, while outsourcing specific tasks such as mapping, inventory and environmental assessment (see Table 2.6).

In some countries, preparation of management plans for state forests is contracted out – for example, in Poland (Landell-Mills and Ford, 1999). However, the power and duty to ensure that management plans are appropriate

Table 2.6 *Forestry activities outsourced by state plantation enterprises*

	Australia	Chile	China	India	New Zealand	South Africa	UK
Inventory		n/a			n/a		✓
Management planning		n/a			n/a		
Establishment and tending	✓	n/a	✓	✓	n/a	✓	✓
Protection/guarding		n/a	✓	✓	n/a		
Harvesting	✓	n/a	✓	✓	n/a	✓	✓
Transport	✓	n/a	✓	✓	n/a	✓	✓

to state agencies' objectives and comply with relevant laws and regulations lie with the agency. Grundy (see Chapter 11) observes that contracting out management of plantations in the UK would achieve little as the assets remain in state ownership and the private-sector managers are not, in general, more competent than the state agency. Where the forest is to be managed for multiple benefits, the management contract becomes complex in an attempt to specify in advance how, for example, production is to be traded against biodiversity over a long period. This is particularly so for the task of transforming an even-aged plantation into a diverse forest, which requires a management plan executed over several decades. In such cases, short-term management contracts are not long enough to show results and a very long contract defeats the object of using the private sector to increase competition.

Securing the benefits of outsourcing
Outsourcing may appear a simple means of securing the benefits of private-sector participation; but effective implementation cannot be taken for granted. Much depends upon the capacity of private actors to deliver services of an acceptable standard. If private contractors are to take over harvesting and management functions, there needs to be a minimum existing level of expertise within the sector. In addition to ensuring that the private sector has adequate skills, a key prerequisite is that the sector is reasonably competitive. Where plantation management skills are concentrated with just one or two firms, anti-competitive behaviour in the sector could undermine any potential efficiency gains from contracting out. Competition needs to be strong enough to ensure that the state gets a good deal. State plantation agencies can help to develop competition by supporting new business start-ups.

The need for capacity-building varies. In the UK, the existence of a substantial state-owned plantation sector helped initially to develop private contracting businesses and provided experienced plantation managers (see Chapter 11). In South Africa, a healthy contracting culture existed before transfer of the state plantations, driven essentially by the private sector. Government has not needed to facilitate this, although most of the large contractors are white run, which is a concern to government who would like to see greater affirmative action in the area (see Chapter 10). But where private actors have never before been involved in plantation activities due to state dominance, capacity-building will be necessary prior to the transfer of any state functions.

Building private-sector networks. Recognizing the importance of private-sector capacity development, the UK's Forestry Commission sponsored the development of the Forestry Contracting Association and continues to fund the Forestry and Arboriculture Safety and Training Council, with the aim of increasing private-sector contracting capacity and the development of technical and health and safety best practices.

Facilitating contractor start-ups. This is mostly likely to occur in countries where private plantations have coexisted beside state-owned operations. Australia's states/territories have used a variety of processes to facilitate outsourcing. Many state agencies invest time in the provision of training and education services for contractors to ensure that contractors are aware of best practice, comply with regulations and are involved in ongoing improvement of skills. In China, state plantation agencies help to develop competition by supporting new business start-ups (e.g. transfers of forest machinery to former state forest enterprise employees; see Chapter 7). In the UK, workers have been encouraged to set up as contractors through contracts tailored to their capacity or financial circumstances (see Chapter 11). In New Zealand, the process used by the New Zealand Forest Service to help wage workers who were being retrenched to set up as contractors to the New Zealand Forestry Corporation provides a useful example of government helping to support the transition from wage work to outsourcing in order to help develop a competitive private industry as rapidly as possible. Workers submitted proposals to set up their own contracting businesses and successful applicants were provided with financial and training assistance, allowing them to purchase equipment and also helping to train them in running their own business.

Maintaining standards with outsourcing. Environmental standards in relation to outsourcing can be established and enforced through contracts; but this is not without its problems. Kun (2000) recounts the programme of transferring management rights of large areas of China's barren lands to individual families to address the shortage of forest lands for fuelwood and other on-farm needs and to encourage farmers to afforest and protect barren hills. Results were mixed; in many instances, farmers continued to neglect forest areas because they feared a reversal of the policy. In the UK, the standards demanded of contractors are clearly specified in Forest Enterprise's quality standards for establishment and harvesting operations.

The state also needs adequate capacity. Even where the state's performance in plantation management has been poor, this does not mean it will have the skills to oversee effective implementation by outsiders. Retraining is essential to ensure smooth transfers of work.

Well-designed contract technical specifications. Contracted outputs (hectares planted, cubic metres felled and extracted, number of maps prepared) need to be SMART (Specific, Measureable, Achievable, Realistic and Time-bound). Quality standards, as well as output quantities, need to be clearly specified.

'Corporatization': Improving efficiency and profitability of state enterprises

Commercialization and corporatization are relatively common. In a review of 23 countries' experiences with privatization, the International Institute for Environment and Development (IIED) found that almost half had completed or initiated some form of corporatization process (Landell-Mills and Ford, 1999). This picture is reflected in the seven cases reviewed for this book. There are different models of corporatization – state-owned company, state agency with the status of a legal person, state agency without separate legal identity – and there is different emphasis on delivery of profit and environmental and social services. In Australia, all states with the exception of Victoria and the Australian Capital Territory have corporatized forest authority activities. South Africa corporatized its plantation management through the South African Forestry Corporation (SAFCOL). The UK adopted the corporatization model as an alternative to privatization of state-owned forests. Its experience with setting up its state forest management agency, the Forest Enterprise, is outlined in Box 2.11.

In New Zealand, corporatization was a stepping stone to privatization of the New Zealand Forest Service's estate. The service's commercial operations were removed to a new state-owned limited liability company, the New Zealand Forest Service (NZFS), which introduced changes to labour agreements, log pricing and business structure. In its first two years of operation, the NZFC generated a cash surplus of NZ\$53 million compared with a NZ\$117 deficit on commercial activities in the final two years of the NZFS (Birchfield and Grant, 1993). At the same time the government made a number of changes in the market and regulatory environment within which the new corporation began to operate:

- making log prices competitive prior to and during corporatization;
- deregulation of the transport industry, which significantly reduced the costs of transporting logs and processed wood products;
- deregulation of the ports, which also reduced costs and the amount of time to export logs and wood products;
- deregulation of the labour market, enabling flexible contracts which reduced labour costs and allowed more flexible working hours;
- deregulation of the financial market, enabling easier access to finance for many businesses, including some plantation businesses, particularly processors needing to restructure and become more competitive.

Corporatization, accompanied by the adoption of commercial business practice, has proved to be a relatively painless way of securing increased effectiveness and efficiency in state plantation management, reducing state organizations' call on public funds. Employees have mixed feelings about it: some seeing it as an opportunity (especially those in senior management positions); others seeing it as threat because it might be a step towards

Box 2.11 Stepping back from privatization to corporatization in the UK

The UK looked at privatization during the 1980s and stepped back. Opposition came from forest users, environmental NGOs, as well as the wood-based industry. There was much concern about the likelihood of loss of access, highly valued in densely populated areas. Furthermore, the Forestry Commission had made much more progress in increasing the biodiversity of plantations. This was partly because they were older, and thus easier to diversify at the stage of felling and regeneration through changing species, age structure and layout. But the Forestry Commission had also explicitly adopted wildlife and landscape enhancement as targets and was making evident progress, whereas the environmental movement feared that the profit motive would dominate private-sector management. It expected it to be easier to apply pressure on a state agency than on a large number of private owners. The wood-processing industry preferred the devil it knew for the reasons set out above. There was, therefore, a large political opposition to privatization, set against the practical difficulties of a wholesale disposal of the state-owned plantations.

The outcome of the review, and subsequent developments, show the government achieving some of its broader economic objectives, while maintaining the provision of non-marketed benefits:

- Regulation was separated from forest operations within the Forestry Commission by setting up separate agencies (the Forestry Authority as regulator and the Forest Enterprise), with clearly distinguished roles and objectives.
- Efficiency was pursued through setting financial and economic objectives for the Forest Enterprise, the agency charged with managing the state forest estate.
- The programme of rationalization through sale of plantations costly to manage because of their size and/or location, and not highly valued for other benefits, such as recreation or biodiversity, was extended.
- A new scheme to encourage private owners to provide access was introduced. This complemented the earlier changes in the grants for planting to promote broadleaves, native woodlands, diversification of species and greater attention to landscape impacts.
- The reorganization of the Forestry Commission was followed by a review of its structure that led to a reduction in its staff numbers and reduced costs.

Source: Grundy, 2002

privatization and an even greater focus on efficiency and profit. Or they anticipate job shedding as part of the corporatization process. NGOs have generally welcomed corporatization (subject to sufficient safeguards) as a way of clearly (or more clearly) separating government's regulatory and management functions.

Transfer of use rights: Widening access to forest goods and services

Governments that have opted to transfer use rights have done so for a variety of reasons. It is an obvious means of increasing access to use rights as a pathway for improving rural livelihoods or to reduce inequalities in the distribution of forest goods and services. It is also being used to engage communities in forest protection. It might be chosen where state enterprises lack the managerial capacity to operate a sophisticated outsourcing system; while they must still monitor compliance with licence conditions, the burden is likely to be lower than that for outsourcing. There may be reasons for the state needing to retain ultimate control through continued ownership of the resource – for example, to protect particular forest values.

Governments do not have a completely free hand. Their options are constrained by their rights over the assets in, and under, their plantations. In the course of developing their estates, governments have acquired an array of different rights, from ownership in perpetuity of all the land's biological and physical assets to rights that are limited to growing trees and to the activities necessary for this purpose (see Table 2.7).

Australia: Ownership and management largely retained in state hands

Australia's states and territories have – so far – generally chosen to retain state control over the management and exploitation of plantations. Their decisions largely reflect the concerns of many Australians about the potentially adverse effects of privatization more generally; the regional economy and, thus, the political significance of the established plantation forests and the processing industries dependent upon them; and the political difficulties inherent in selling state land on a large scale, particularly to non-Australian investors (see Chapter 5). The only case of full privatization of a state-owned plantation resource in Australia to date has been the sale of the Victorian Plantation Corporation (ViPC) to Hancock Victorian Plantations Pty Ltd, a subsidiary of the US-based Hancock Timber Resource Group, in 1998, although others are mooted.

The ViPC privatization was achieved in four stages:

1   the establishment of the ViPC in 1993, under which the plantation business was treated as a sole entity and on a commercial basis rather than as a government department that had multiple facets and competing budgetary obligations;

Table 2.7 *An array of state ownership rights over plantation assets*

Asset	State's ownership of the asset
Trees	❑ Exclusive rights to trees growing on the land ❑ Rights subject to a profit-sharing agreement with the landholder
Land	❑ Exclusive and unrestricted use of the land ❑ Exclusive use of the land, but use restricted by clauses in freehold title or lease ❑ Shared rights to use the land ❑ Ownership rights subject to land claims
Plants and animals	❑ Exclusive rights to all species ❑ Exclusive rights only to certain species ❑ No exclusive rights ❑ Rights vested with another owner ❑ Rights are public
Water	❑ Exclusive rights to use rivers and lakes in the plantation ❑ Shared rights ❑ Rights vested with another owner ❑ Rights are public
Minerals	❑ Exclusive rights to minerals ❑ Shared rights to minerals ❑ No rights to minerals
Access	❑ Exclusive right to determine who may and may not have access to the forest ❑ Joint rights to have and grant access ❑ Rights limited to employees of the state enterprise for the purpose of forest management ❑ Public has statutory right of access on foot

2   the commercialization of longstanding wood-supply agreements, most of which were enshrined in acts of parliament;
3   surveying of plantation lands vested in the ViPC by accurate delineation;
4   international offer for sale which was managed by the Victorian Treasury and international consultants.

The ViPC was sold as a business entity rather than as a share float, as has been the case with some other privatizations in the country. This option was chosen because it was anticipated that it would generate the greatest return to the vendor (the state), and because it was thought that it would foster retention of the whole estate under single ownership. It largely achieved the latter intention; but the former contention cannot be tested.

China: Granting long-term use rights to households

With the extension of the agriculture sector's household responsibility system to forestry during the early 1980s, China has seen a rapid and significant shift in forest use rights from collectives and, more recently, state-owned forest enterprises to individual households. Rights transfers have gone furthest in the southern collective regions. By 1986, an estimated 69 per cent of forest land in eight provinces had been handed to households for use and management (Lu Wenming et al, 2002). Use rights have also been transferred by state-owned forest enterprises, most often to employees. Land is frequently offered as a substitute for salaries and pensions, as well as for severance pay for workers who have been dismissed. These payments are called 'salary land', 'pension land' and 'employment land', respectively (CAFLU, 1997; Lu Wenming et al, 2002). The transfers have generated a number of benefits:

- increased labour and capital input in forestry by rural households;
- new employment opportunities;
- diversification in the income base of rural households;
- improved watershed protection and soil fertility.

India: Granting rights to community groups under Joint Forest Management

India's Joint Forest Management (JFM) programme could be considered outsourcing rather than a transfer of use rights. Under JFM, community groups protect and manage forest patches adjoining villages. By March 2003, over 63,000 JFM groups were protecting and managing over 14 million hectares of state forests, including many plantation areas. The communities are generally given a share in income from the forest or plantation – for example, community groups protecting cashew plantations in West Bengal are entitled to 25 per cent of the profits from the sale of cashew. Reported benefits of JFM include (Khare et al, 2000):

- restoring degraded forests;
- increasing the availability of fodder grass;
- sharing pole harvests;
- sharing the income from forest products.

JFM has provided many village communities with greater legitimate access to an important livelihood resource. Several JFM groups have been able to create village funds, enabling them to undertake other development activities. However, over-emphasis on commercial timber production has put a question mark against JFM's ability to deliver its original social and environmental objectives unless changes are made in the institutional and benefit-sharing arrangements (Khare et al, 2000; Saigal et al, 2002). The challenges faced by JFM are outlined in Box 2.12.

Box 2.12 Joint Forest Management: Work in progress in transferring forests to communities in India

In India, Joint Forest Management (JFM) has had many positive impacts (Saigal et al, 2002); but many challenges still remain. In most states, JFM is still depends upon donor-funded projects and its long-term viability is yet to be established. In most states, JFM programmes have been established on the basis of mere administrative orders that can be changed or withdrawn by the forest departments (FDs) at any time. There is no tenurial security for the participating communities. The share of communities in the revenue from forests is still low in many states and there are restrictions on collection and sale of several commercially valuable non-timber forest products (NTFPs). A key challenge is the right link between *Panchayati Raj* institutions (PRIs) and forest protection committees (FPCs). There is considerable confusion over the role that PRIs should play in JFM. One suggestion is the FPCs should work under PRIs; but many fear that it will destroy the FPCs as PRIs are often highly politicized and large bodies, which are not suited for JFM.

Khare et al (2000) observe that arguments rage about who is paying for and who is benefiting from JFM, and whether good forestry is being practised. They conclude that JFM is a policy with feeble roots. The main JFM policy statement can, in theory, be withdrawn at any time in the absence of greater legal recognition. Many of the initial state JFM orders were pushed through by *ad hoc* initiatives taken by interested individuals without any open debate or discussions. As a result, many JFM orders continue to be riddled with serious contradictions. Some state that JFM orders have been issued under pressure from donor agencies who started demanding inclusion of JFM as an important component of large externally aided forestry projects.

They note four institutional challenges:

1   Government institutions have different political compulsions and different pressures on their budgets from central government.
2   Village organizations in most states have no autonomous status and can be dissolved by the forest department.
3   Transfer of decision-making authority to local users implies a corresponding reduction in power of the forestry department, which continues to be resisted by forestry bureaucrats.
4   Radical changes in forestry department organisation, culture and skills are required.

New Zealand: Leading the way in granting long-term, exclusive rights

Following the decision to privatize the New Zealand Forestry Corporation (NZFC) after its corporatization, the government faced three further decisions:

* *Whether to sell the assets or the business.* The government decided to sell NZFC's assets rather than sell NZFC as a going concern. Birchfield and Grant (1993) give the reasons:

    > *...there was little or no logic to the corporation's portfolio of forests and processing activities and it would take a considerable time to rationalize the business to the point of it being a credible contender for public listing. Now that sizeable cash surpluses were establishing a value [for NZFC's assets], it [was] more logical to sell the forests to the highest bidders.*

In addition, there was uncertainty as to whether the New Zealand market was in a position to absorb a float the size of the NZFC. Finally, the uncertainty arising from the existence of Maori land claims for much of the land on which state plantations were established meant that selling the NZFC as a going concern would likely result in a court action.

* *Whether to sell both the land and the trees or only the trees.* Selling the land was unlikely to be acceptable because of ongoing Maori claims and strong public resistance to foreign ownership of land. The decision was therefore made to sell the trees but not the land by selling cutting and management rights in the form of long-term leases called Crown Forestry Licences (CFLs).
* *How to sell the assets.* NZFC organized and was the sales agent for the plantation sales. NZFC's role created some concern amongst some members of the New Zealand forestry industry who believed NZFC had motivation to delay sales, having just invested considerable effort in creating and profitably running the NZFC. The NZFC estate was split into 90 parcels, each an appropriate unit for forest management, ranging in size from 51 hectares to 132,112 hectares, in theory allowing maximum flexibility by allowing small and large organizations to bid. Each unit had a CFL assigned to it with individual terms and conditions of sale. CFLs were designed to be as close to freehold land with rent as possible in order to minimize the constraints in CFL holders. Identifying rights over parcels took a long time but was essential in order avoid future problems. Rights were written into the CFLs as covenants (e.g. vehicular access rights, conservation requirements and grazing rights). All CFLs contained a 'wander at will' clause allowing pedestrian access unless safety or other reasons precluded it. This wander at will clause was annulled if the land was transferred to Maori or other private landowners. All CFLs could be terminated in the event of the land being transferred to Maori owners. Once a termination was given, the land would be returned as the trees were harvested. Some CFLs had an initial fixed term before termination notice could be given to allow for adequate infrastructure to be developed or for other conditions to be sorted out (e.g. for long-term contracts to expire).

The transfers raised a number of issues. First, the provisions put in place until such time as the land underlying the plantation was transferred from the Crown to Maori owners, which included the rights to give access through plantation roads, would also transfer to Maori (i.e. access rights over the roads were not guaranteed to the CFL holder once ownership of the land transferred). Lack of such a guarantee is currently creating concern surrounding claims over the 22 CFLs granted over the Central North Island plantations (Neilson, 2003).

Another issue was whether CFL holders should be required in perpetuity to replant the land after harvesting. In the initial round of sales in 1990, no replanting conditions were imposed. By the time of the 1992 round a condition was imposed that the land must be replanted or converted to another sustainable use that had to be approved by government in order to reassure a range of groups who were concerned that replanting might not take place (Clarke, 1999).

The question of processing also became a consideration. During the 1990–1991 sales, bidders were not required to guarantee to maintain or expand domestic processing. In 1992, bidders were required to honour five-year supply agreements with existing clients, and in 1996 bidders were required to demonstrate their intention to add value to the resource within New Zealand (Clarke, 1999).

South Africa: Building on New Zealand's experience, but for a different mix of objectives

The starting point for South Africa's privatization of plantation management and exploitation was the policy of withdrawal by the state from direct involvement in commercial forestry plantations. Instead, the government would create the conditions and policies necessary for others to manage forests in the national interest and regulate their actions (Mayers et al, 2001). The policy mirrors a wider national programme of economic reform and liberalization in post-apartheid South Africa, which is driving privatization initiatives across several sectors. These include transport, manufacturing industries (including defence), telecoms and public-sector infrastructure development and management (including roads and hospitals). Privatizing forests is therefore seen politically as one of several such processes taking place within this context of economic reform. In common with New Zealand, the basis around which forestry privatization has been formed has included:

- A belief that privatization offers opportunities to attract investment and expertise needed to revitalize assets which are perceived to suffer from chronic under-investment. The reality, particularly in terms of the homeland plantations, is that they have not suffered from a lack of resources, but rather an institutional inability to spend the resources efficiently to achieve economic sustainability and growth.
- A fiscal imperative to reduce the burden on the state of subsidizing inefficient (relative to the private sector) or even loss-making government plantations. While SAFCOL was able to make an operational profit within

two years of its creation, it is argued that there are many efficiency gains still to be achieved. The Homeland plantations, however, have historically been managed at a significant loss to the state.

- A recognition that continuing to manage plantations potentially conflicts with the performance of government's regulatory role. While there can be an institutional separation between regulation and operation within the state, the reality is that the institutional importance of one function over another is determined by budgetary allocation and organizational structure. If government's forestry functions are weighted heavily in favour of its operational responsibilities, there is a tendency to see 'small' regulation and policy-making functions as 'add-ons' to the organization, rather than its core.

In addition to these commonly occurring privatization objectives, the South African government was also aware that privatization offered a unique opportunity to achieve a number of its other White Paper policy objectives in a meaningful way. These included the need to:

- Achieve a wider, more representative, pattern of ownership in the forest sector, particularly amongst South Africa's historically excluded groups.
- Consolidate forest resources often artificially split along former Homeland boundaries, a separation which resulted in inefficiencies and distortions in resource use to the detriment of the industry's rational development.
- Improve efficiency in processing industries by increasing effective competition for raw material supplies. South Africa's saw-milling industry has been characterized by long-term structural inefficiencies, resulting in large part from the distorting influence of government as a major forest owner.
- Formally recognize the land, access and use rights of rural communities, many of whom were dispossessed of these rights when these plantations were established.
- Facilitate black empowerment through direct participation in forestry operations, training and skills development, affirmative action in management, and entrepreneurial opportunities through outsourcing, partnerships, procurement and easier access to financing.

The opportunities surrounding exclusion and black empowerment shaped South Africa's privatization in a fundamentally different way from that of New Zealand. Conditions built into the bidding process ensured a minimum degree of involvement by black entrepreneurs.

UK: Limited success with transferring rights to communities
In the UK, the programme of limited asset sales that began during the 1980s sparked interest among rural communities in purchasing state forests; but progress has been much slower than in some of the other countries studied, particularly China, India and South Africa. The Forest Enterprise

found it difficult to reconcile communities' aspirations with the commercial management of the estate and a culture of 'ownership' and control over the resource. The problem was compounded by lack of understanding of what communities wanted – some communities having unrealistic expectations – and a lack of the skills needed for engaging with communities. The problem was highlighted at the Scottish village of Laggan where there was uncertainty over the future of the local forest, which had been a candidate for sale. The Forest Enterprise had withdrawn employment from the area in the past and was increasingly using contractors. Lack of local contact and the fears over disposal triggered the community to try to acquire the forest. After protracted and complex negotiations over a number of years, a forest management partnership between the community and the Forest Enterprise was established in 1997. As a result of the experience gained from cases such as Laggan, the Forest Enterprise has made a radical change to its approach to local communities who want to be involved in the management of Forestry Commission forests; but genuine transfers of use rights have occurred on only a small scale so far. Action on forests in the UK contrasts markedly with the imaginative policy of the Scottish government of supporting land purchases by crofting and other rural communities (Garforth and Dudley, 2003).

Total divestment of the state's interests: A rare choice

The transfer of exclusive rights to all of the assets vested in state agencies is rare. Of the seven countries studied, Chile is the only one to have transferred all of the state's plantation assets to the private sector: in some cases to the landowners with whom the state agencies had entered into partnerships, in other cases to private corporations that had no previous rights to the land. Driven by the liberal market philosophy of the government of the time, no outcomes were sought other than to transfer all commercial activity to the private sector.

Key reasons governments may choose to transfer ownership rights include:

- *Efficiency.* A belief exists that partial rights transfers (e.g. of use or management rights) are less effective in stimulating entrepreneurial drive and improved efficiency amongst private actors (see the discussion on efficiency in 'Changing ownership and management: Some global patterns' in Chapter 1).
- *Budgetary requirements.* The government's plantation authority may be severely short of cash and unable to continue to perform its responsibilities. Privatization transfers its liabilities and cashes in on its plantation assets.
- *Limited environmental and social concerns.* Full ownership privatization is likely to be favoured in plantation areas where the public benefits attached to the plantations and, thus, public resistance to their transfer are relatively small.

Challenges of use-rights transfers

Use-rights transfers pose a number of unique challenges for governments, varying in degree with the nature of the rights concerned and the context (political, social, economic and environmental) within which the transfers are made:

- Getting target groups, and especially communities, to engage in ownership and management in such a way as to attain the benefits of transfer (see India's experience with Joint Forest Management in Box 2.12).
- Ensuring best value for money in terms of cash, labour or other payments made in return for the acquired rights, and continuing subsidies that might be necessary.
- Getting the new rights holders to achieve and maintain satisfactory standards of management.
- Managing job shedding and other social impacts of change.
- Balancing the rights of the beneficiaries of transfers with the needs and expectations of other stakeholders.
- Overcoming resistance in state enterprises stemming from fears of power and job losses.

In Chapter 3 we draw from the experience of the case study countries and wider literature to suggest ways of approaching these challenges, and offer for guidance 'best-bet' practical instruments and processes for achieving successful transfers.

Balancing Acts: Reconciling Public Policy and Transfer of Ownership and Management

Mike Garforth, Natasha Landell-Mills and James Mayers

Increasing private-sector participation in state forest plantations can bring many benefits. In its simplest form – outsourcing – state plantation enterprises can reduce costs and increase profitability, bringing savings to hard-pressed budgets or enabling increased expenditure on supplying the public goods and services that plantations can provide. Outsourcing can help to develop capacity, preparing private actors for taking on the responsibilities that accompany rights to develop and exploit plantation resources. This is the point at which substantial benefits begin to flow to the state (through the realization of capital assets) and to rural people (through new livelihood pathways). For this reason we focus in this chapter exclusively on the transfer of ownership and management from the state to private actors.

Opportunities demonstrated by experience

The experience of the case study countries shows us that transferring rights to exploit plantation assets offers a variety of opportunities. Some of them might be considered generally applicable, although their impact will depend upon local contexts. Others are context specific, occurring only in particular circumstances that are related to political or social history.

Demonstration of political philosophy. Plantation privatization, or moves towards it, may have been driven largely by political philosophy, untailored to the specifics of the task at hand, as in Chile under the rightist governments of the 1970s and in the UK under the Thatcher government of the 1980s. This may work on its own terms, but does little for either plantations or the people who need them. Where the governance framework allows stakeholders an effective voice, 'privatization for privatization's' sake is likely to fail unless real benefits can be demonstrated.

Increased operational efficiency and profitability of asset use, leading to gains to the national economy. Although the higher entrepreneurial drive of the private sector can be expected to exploit plantation assets more efficiently than state organizations, evidence for this from the case studies is limited. In New Zealand's case, although some commentators believe privatization has led to increased efficiency and competitiveness, particularly in global markets, others believe that it is not possible to separate the impacts of market, labour, port and transport deregulation from the impacts of corporatization and privatization. Others have questioned whether corporatization and privatization could have achieved improved economic outcomes without the associated economic restructuring (Easton, 1994, cited in Hall, 1997; Clarke, 1999). The New Zealand Forest Service (NZFS), which was only sold in 1996, operated as a state-owned enterprise for almost six years and was generally considered a very successful commercial plantation company. There is also evidence from New Zealand of short-term decreases in economic performance due to privatization and associated changes (Grebner and Amacher, 2000).

Increased investment in wood processing. Here, too, the evidence is limited. The private sector has shown its willingness in the past to invest in processing capacity without the added security of wood supply that might come with ownership of long-term exploitation rights. New capacity is being developed in Australia and the UK, where control over much of the raw material supply remains in state hands. New Zealand has experienced increased investment in domestic processing since privatization. Clarke (1999) found that 'All new forest owners have invested, or intend to invest, in value-added processing... Of the NZ$1600 million of intended investments in the period 1990 to 2005, 90 per cent is attributable to the purchasers of state forest assets.' Hall (1997) attributes much of this to the success of privatization in attracting foreign investment capital into New Zealand. Some question whether investment in processing may, in fact, have been curtailed as capital had to be tied up in buying trees, leaving less available to invest in processing facilities. This raises the question of whether using a different model, such as selling the cutting rights, might have resulted in more processing investment; using a cutting rights model would no doubt have raised its own set of issues (see Chapter 9).

Healthier state budgets. Opportunities to reduce public spending depend upon the nature of the rights and responsibilities that are transferred. If a state-run enterprise is genuinely inefficient, transferring the enterprise's activities to private actors as part of a use rights agreement will result in savings that should be reflected in the price paid for those rights. We can distinguish two types of inefficiency. First, the scope or volume of the activities carried out by the enterprise may be wider or higher than is necessary to achieve the government's objectives. Such a situation can arise where a state enterprise has freedom to 'set its own agenda'; therefore, while delivering on the objectives given to it, the enterprise engages in 'fringe' activities. The second type of inefficiency occurs when unit costs of activities are inflated, perhaps because of high labour inputs and the extent of the government's post-transaction responsibilities. Based on the experience of the countries studied, it seems likely that both types of inefficiency can be fully addressed through corporatization (to focus the enterprise on the activities that matter) and outsourcing (to bring down unit costs). Net savings to the public purse will be affected by the continuing, possibly increased, investment that the government will have to make in ensuring that private actors deliver the outcomes that are built into transfer agreements and/or comply with forestry-related laws.

Redirected state roles. Transfer of exploitation rights together with management responsibility can be a key element in reforming the role of the state. By removing themselves from any commercial interest in plantations, state organizations can concentrate on the 'core business' of government, creating an environment that enables private actors to deliver public-policy objectives. Resources freed up in this way can be used to increase investment in other roles, such as systems of practical regulation, extension, research, and monitoring and evaluation. However, redirecting the role of the state can be achieved to a substantial degree by separating commercial and strictly government functions within a government framework as, for example, in Australia and the UK, and in New Zealand where the separation was made several years before the decision was taken to 'sell off' the state-owned plantations. Eastern European countries that have taken steps to restructure state forestry institutions have also generally followed the path of separation within government, adopting different models of state-owned forestry enterprise.

Poverty reduction and improved rural livelihoods. China's experience of granting long-term use rights to land for forestry shows that substantial gains can be made if the transfer is well designed and is based on a solid understanding of the links between forests, forest enterprise and local livelihoods. India's experience with Joint Forest Management (JFM), while promising much, demonstrates the importance of transferring rights sufficient to translate into livelihood benefits and establishing institutional systems capable of pushing through the objective. Establishing transaction rules that target poor groups and prevent capture by elite groups will be particularly important where social and economic power structures work against transfer processes, as in South Africa. The evidence from the case studies suggests that the opportunities are greatest in developing countries; but this might be simply because developed countries have not been imaginative or determined enough in using transfers as a mechanism for improving rural livelihoods.

Empowerment of disadvantaged groups. South Africa is the only country to have aimed transfers explicitly at empowerment of disadvantaged groups. While it is still too early to gauge success, experience there can teach us useful lessons on how to shape the process, if this is the objective.

Righting past wrongs. In specific contexts transfers can redress cases of past land and resource appropriations by the state that are now deemed to have been illegitimate, as in Eastern European countries. Elsewhere, as in South

Africa, forest transfers have been taking place within a wider programme of land and social reform.

These last three opportunities present the greatest challenges to the designers of transfer processes because they require targeting of specific groups who are likely to lack the capacity to exploit their new rights and exercise the responsibilities that come with them, and more careful management of trade-offs – for example, between offering secure tenure rights for bidders versus protecting underlying land ownership and informal tenure rights. The political imperative may be so strong that trade-offs might be ignored. We have seen this with the restitution of forest lands in countries of the former Soviet bloc, where little attention appears to have been paid to ensuring that the forests involved would continue to be managed responsibly.

Key challenges

Engaging local groups in ownership and management

Dilemmas in local participation. For governments concerned with the social impacts of privatization in rural areas, a key challenge is ensuring local participation in both the process and the final outcome. Small local landholders, however, face a number of hurdles in trying to take part. Common constraints revolve around their lack of capacity to get involved (e.g. requirements for management plans are often complex and alien to what communities are used to) and the lack of political will to truly devolve responsibilities (Matose, 1997, cited in Lindsay, 2000).

'Myth' of community. A key question is how state law can ensure recognition of local authority over resources. One approach is to designate on a uniform basis a local body that would have control over a predefined area – for example, a village council. Another approach is to provide for the recognition of different groups formed around different functions and objectives. For instance, in Nepal the forest law refers to user groups that will have land turned over to them – these are essentially self-defining groups and neither their membership nor the area they manage have anything to do with local government boundaries. The central problem with both approaches is that of 'locating the community'. Even within single villages there are overlapping and often conflicting ideas of community, often bearing little resemblance to what outsiders see or want to see (Lindsay, 2000).

Recognition of legitimate and motivated groups. Flexibility is required with regard to how state law handles the recognition of local groups. The law should not try to squeeze existing local institutions, which have roots in local values and practices, into forms that are too complex and alien to the local situation and then standardize that form across many different social settings. The result could be to create institutions that have little legitimacy among their members (Lindsay, 2000). Fisher (2000) concludes that natural communities are a better basis for collective action than artificially constructed or administratively convenient units (because people are less likely to breach agreements when doing so will interfere with existing social arrangements). He argues that collective action around resource management is more likely to occur when the boundaries of the resource and the boundaries of the social unit managing the forest coincide. Where communities are defined in terms of the formal political and administrative structure, there are real risks that responsibility and authority will be applied to a 'community' level that is inappropriate in terms of what is known about effective collective action.

Secure tenure instruments are essential but not sufficient. Rights on their own are not enough; they must be supported by the capability to claim and defend them. In the case of devolution to communities, this is likely to mean defence against more powerful actors. Clear constitutional guarantees, as well as specific supportive legislation and regulations, are necessary but are still not sufficient on their own. Poor people need to be aware of their rights and to be able to access effective routes to recourse. Devolution of forest use and management rights requires effort to establish the effectiveness, legitimacy and accountability of local institutions. Management of budgets, costs and benefits should be devolved along with responsibilities. Local institutions need sufficient autonomy to act on, modify and enforce local rules. In addition, as well as setting guidelines for the protection of legitimate public interests, regional or national laws should define rules according to which communities interact with outsiders and which provide basic protection for individuals and disadvantaged groups against the abuse of local power (Mayers and Vermeulen, 2002b).

Disadvantages of some formal political structures. There is a tendency for bureaucracies to focus on formal political structures because this makes it easier to identify representatives, a clear legal basis exists, and it is procedurally simple. But there are major disadvantages (Fisher 2000):

* The formal political system is often adversarial, whereas resource management requires consensus.
* The representatives do not represent all interests, so there is likely to be limited adherence to decisions.
* Formal political systems tend not to coincide with 'natural' user groups.
* Collective action is inhibited if relationships are not 'embedded'.

Conflicts between arms of government. The picture can be even further confused where different parts of government recognize different local-level groupings. In Nepal, the Forest User Group (FUG) system has been quite successful in transferring responsibilities to local levels. The Forest Act gives authority to the Department of Forestry to hand over forest land to a FUG, which then becomes the overall decision-maker and manager of the designated forest area. But the Decentralization Act provides village development committees (VDCs) with the authority and responsibility to form user committees, which then decide how to manage the resources within the identified VDC boundary. Due to overlaps in both acts, conflicts between FUGs and VDCs and between the respective government departments that back them up have been increasing (Singh, 2000; Upreti and Shrestha, 2000).

Blending strengths of local government and peoples' organizations. Effective decentralization or devolution therefore requires devolved decision-making and devolved power and authority. The process needs to informed by an understanding of the social basis of collective action and must be consistent. Authorities need to think of ways to combine the concerns of local governments with those of other non-formal groups (Fisher, 2000). Lai et al (2000) also observe that strong peoples' organizations are the keys to successful community-based forest management and note some problems from experience in Indonesia:

- Peoples' organizations often lack the organizational and technical capacity to properly manage commercial aspects related to community-based forest management agreements (see Box 3.1).
- Many communities lack working capital and have little or no previous financial management experience.
- Peoples' organizations have difficulty negotiating fair market prices, finding affordable transport, arranging payments, assuring quality standards and meeting pre-payment requirements.
- Government field offices are usually unable to provide all of the assistance needed by forest communities, especially with regard to cooperative business management.

Box 3.1 South Africa: Local enterprise is motivated, but needs support

Experience from South Africa is that communities and small enterprises can organize themselves quickly when the benefits have been presented to them clearly; but they will need short-term assistance in the form of access to experts, such as lawyers and transaction advisers, or for business management training. For long-term sustainability and human resources development, institutional support can be provided by a well-established private-sector partner within the consortium. Sound institutional arrangements and provision for institutional development and support (once community-based legal entities have been established) are key to ensuring that they play a developmental role and that trustees remain accountable to their members on management and the distribution of benefits.

Source: Dlomo and Pitcher, 2002

Ensuring best value for money

Administrative allocation versus auctions. Auctions are generally recognized as maximizing revenue from the sale of land use rights. They can also counter problems of inefficient administrative land allocation and associated forest fragmentation by allocating use rights to the highest (and implicitly, though not always, the most efficient bidder; Lu Wenming et al, 2002). Box 3.2 summarizes the experience of auctioning use rights in Lushan County, China.

Box 3.2 Experience of auctions in China

Lushan County has been a local leader in introducing auctions for forestry rights in Sichuan Province. As in other counties, the early years of de-collectivization involved the administrative allocation of forest land to local households. In general, recipients paid a flat annual rent or a share of the profits. During the early 1990s, auctions were increasingly used to allocate land. Auctions involved collectives calling for bids from potential buyers in the form of up-front payment or annual rentals.

Use rights to about 5000ha of land have been auctioned between 1993–1998. Fifty per cent of the land was allocated for 10 to 40 years. Longer term rights of between 70 to 100 years were also allocated, but only in about 2 per cent of cases and concentrated in remote mountainous area.

Since being allocated, a healthy trade for forest land rights has evolved. Sellers include co-operatives, town governments and local forest farms. Buyers include farmers, non-farming individuals, co-operatives and private companies.

Source: Lu Wenming et al, 2002

Recurring needs for subsidies. An important factor in weighing value for money is the intrinsic profitability of commercial forestry. If it is lower than the usual rate expected from long-term investments, the interest of private sector will be small and may be only from those wishing to asset-strip with a quick exit. The government may have to promise to keep paying subsidies – promises that are usually discounted. The best approach will then be to seek out potential owners with a direct interest in wood production and other plantation products and services (see Chapter 11).

Bid price is not the sole criterion. South Africa has used the bidding process to enable the government to select the preferred bidder whose bid best reflects its multiple objectives. This need not be the bid with the highest price. Bidders were presented with government's objectives and were required to submit bids against a set of qualitative and quantitative criteria. While the range of bid criteria potentially makes for a complex bid evaluation, the approach represented an efficient market-based means of striking a balance between public- and private-sector objectives (see Chapter 10).

Enforcing standards after transfer

Maintaining standards with outsourcing. Environmental standards in relation to outsourcing can be established and enforced through contracts; but this is not without its problems. Kun (2000) recounts the programme of transferring management rights of large areas of China's barren lands to individual families in order to address the shortage of forest lands for fuelwood and other on-farm needs and to encourage farmers to afforest and protect barren hills. Results were mixed; in many instances, farmers continued to neglect forest areas because they feared a reversal of the policy.

To include or not to include blocks of natural forest. One specific environmental issue that remains unsatisfactorily concluded in South Africa is the management arrangements for blocks of indigenous forests that are closely associated with plantations. At the time of determining the lease area, it was decided to exclude out any block of indigenous forest that was large enough to be considered its own management unit. All other smaller areas were left in the lease to be the responsibility of the tenant. It would appear that the criteria used at the time might have resulted in a number of indigenous forest areas being excluded from leases which should have been left in. Negotiations are now taking place with tenants and prospective tenants to either have these areas put back into the leases or to manage them on contract through side agreements to the leases (see Chapter 10).

Improving environmental values is costly. With the exception of South Africa, the countries studied have relied on the broad governance system to enforce environmental standards after transfer of assets. The UK experience is that a system of controls on felling, planting and replanting, with grants to compensate for low returns and/or negative impacts on profitability, can ensure that private forests are managed responsibly and without environmental damage (see Chapter 11). However, the costs of this type of system are substantial. Merely maintaining the level of non-market services may not be good enough. The aim may be, as in the UK, for increased biodiversity and social services. While a system of controls, grants and consultation can ensure responsible forest management, moving private owners to the next stage – an imaginative development of plantations to increase public benefits – is more problematic.

Using certification as a proxy for government monitoring. The new owners of South African former state plantations are required to achieve certification against approved standards within two years of commencement of the lease. National standards for sustainable forest management, required under the National Forests Act, are in the process of being developed. Certification is accepted as a neat, cheap option for government which is likely to give as good an impression of performance management against sustainability criteria as anything that government could set up for such a purpose. There is, however, a growing discomfort around the Forest Stewardship Council (FSC), in particular, because it is seen to be constantly 'raising the bar' or 'moving the goalposts'. It is very unlikely, for example, that any small-scale operator on a category B plantation could get FSC certification in terms of how this is currently perceived. What is needed in South Africa is a more tailor-made solution to deal with the smaller-scale, less well-resourced plantations and their managers (see Chapter 10).

Using organizational statutes as a proxy for government monitoring. In the UK, concerns about loss of public benefits led the government to introduce arrangements to give first option to purchase forests with high nature conservation, recreation, cultural or heritage values to organizations mandated by their statutes to manage for the public good. Under these arrangements sales have been made to non-governmental organizations (NGOs), government nature conservation agencies and local authorities. The option of building safeguards into the title deeds was rejected because it was not certain that a requirement to, for example, maintain provision for public access could be enforced.

Managing social impacts of change

Jobs shed in one sector may lead to opportunities in another. Change aimed at achieving greater efficiency, as with outsourcing and transfers of state assets to more efficient owners and managers, will be likely to lead to job losses from the sector. Those who have to leave the sector may find employment in other sectors so that, viewed from a perspective of national social welfare, there could be a net gain after taking into account the greater efficiency and profitability of plantations and the recruitment to other profitable enterprises outside forestry; but this will not always be the case. Those who remain in the sector may be faced with poorer conditions of service.

Potential impacts on employment. Job losses are inevitable when responsibility for plantation exploitation and management is transferred from overstaffed state enterprises to private actors, although the impact of transfer may be lessened when the state enterprise has been through a process of corporatization and commercialization. The threat of job losses can lead to resistance strong enough to delay or even block transfers. Some of the actions that governments can take to mitigate the impacts of transfers on employment are illustrated by South Africa's experience (see Box 3.3), though as Dlomo and Pitcher observe, there is still much to do to manage the short-term negative impact of asset transfers in spite of these safeguards (see Chapter 10). Additional long-term support is needed to assist redundant workers to maintain their activities sustainably into the future. The problems have been exacerbated by the fact that most of the workers who have left government employment during the privatization were old, low-skilled and close to retirement. They were offered retraining options and business skills training within the framework of a 'social plan' with a very limited budget that was not implemented very imaginatively. Most workers therefore did not get a great deal out of it.

Box 3.3 Managing potential impacts on employment: Lessons from South Africa[1]

South Africa employed the following mechanisms to mitigate the impacts of transfers of category A plantations (those of immediate high commercial value) on employees of the state plantation agency:

- A condition of the transaction, captured in the sale of business agreement, is that all South African Forestry Corporation (SAFCOL) workers will be transferred across to the new employer on existing terms and conditions of employment.
- An industry norm number of Department of Water Affairs and Forestry (DWAF) workers will also transfer across with any former homeland plantation, and they will be taken on at SAFCOL wage rates, and terms and conditions of employment. Any DWAF worker transferring will receive a transfer package to compensate for the difference in salary between what they currently earn in DWAF and the SAFCOL rate. This compensation is calculated over three years and paid as a lump sum at the time of transfer. A variation on this model is expected for the category B plantations (those of lower commercial value).
- Employee Share Ownership Plans (ESOPs) to ensure that employees have a real stake in the plantations are also promoted in the restructuring. Noting that both workers and management contribute to the success of the enterprise, an attempt is being made to ensure that employees, and not just management, will have an equal opportunity to acquire shares. No ESOP has yet been established.
- There will be a moratorium on retrenchment of transferred workers by the new company for three years (again captured in the sale of business agreement).
- Bidders were required to present human resource development plans that addressed affirmative action and effective human resources development planning. This was a criteria used in bid selection.
- Any DWAF worker who could not be transferred was either redeployed to a category B plantation (thereby making the over-staffing there even worse) or they left the public service via a voluntary severance instrument, retirement or early retirement (standard in the public service). A new Public Service Framework Agreement (2002) now allows for employer-initiated retrenchment in the public service; this was not available at the time, and it is likely that this will be implemented in terms of the category B privatization.
- Any worker leaving the public service voluntarily or otherwise is eligible to participate in a social plan. This is currently delivered in the form of a package of training and counselling services to help workers deal financially and psychologically with unemployment, and also to retrain for re-employment or self-employment. Most workers, to date, have elected skills that are suitable for self-employment at a micro or household level.

Source: see Chapter 10

Balancing tenure rights for bidders with other objectives

Weak tenure = low value. In partial or total asset transfers, beneficiaries need secure tenure rights. Weak tenure rights will lead to discounting of the price offered or, in the worst case, no offers being received.

But outright tenure for one = injustice for another. One of the most challenging aspects of the South African privatization process has been to secure reasonable tenure security while protecting communities' underlying land and other informal tenure rights. The solution was to lease the land rather than sell it and to construct the lease to provide the lessee with security of use rights to the land and ownership of the plantation for 70 years. The lessee pays a market rent that is held in a trust fund. When the underlying landowner is identified, any accumulated payments and ongoing rent will be transferred directly to the landowner. The tenant gets full undisturbed possession of the land subject to the requirements of the Forest Act to allow public access for cultural, spiritual and recreational purposes (see Chapter 10).

Overcoming the resistance of government actors

Transfer initiatives require cultural change within state institutions. Resistance within state institutions has slowed transfers of ownership and use rights to private actors and restricted the range of rights that have been transferred. Resistance may occur for various reasons: transfer subjects may represent a major source of government income (Lu Wenming et al, 2002); they confer power on state forest management enterprises; and they provide secure and attractive employment for enterprise staff. Even where state agencies embrace transfer processes, their own agendas may severely limit the benefits. In India's JFM programme, the imbalance in power and control structured into the forest department and in local community institutional relationships is more geared to extending the department's control over the community than to nurture self-governing resource management by the villagers. The arrangement essentially uses the community as an instrument for achieving the forestry department's interpretation of forestry policy objectives (Khare et al, 2000).

Cultural change needs to be driven through all levels of the organization. Statements of intent, policies and guidance issued from the centre will have no impact unless accompanied by training to get buy-in from the forest managers at the 'front line' and to provide them with the skills they need to engage with transfer beneficiaries (Hobley, 2002).

'Best-bet' practical instruments and processes

In this section we provide guidance on how to prepare for and manage transfers, drawing from the experience of the case study countries. We have set out the guidance under four broad headings:

- option selection and design;
- process and change management;
- instruments for ensuring that new owners and managers continue to serve the public interest;
- managing the government's post-transaction responsibilities.

The guidance is aimed at those who can influence decision-making about transfers of ownership and management control for state-owned plantations. Such agents may be in government, the private sector or civil society. They may have explicit powers for key decisions, or they may have the potential to be effective managers or to improve processes, but need guidance to help build their capabilities. Whoever they happen to be, they should keep at the front of their minds two crucial points when considering and, especially, when applying the guidance:

- *Context is all important.* No pronouncements about what to do in one context will quite fit the bill in another; there are no magic bullets, no one size fits all. However, a 'checklist' approach to possible actions, options and ideas on how to proceed – based on experience in contexts where some progress has been made – can stimulate further ideas, learning and the tailoring of ways forward specific to context.
- *Freedom of choice is a rarity.* Key contextual factors that will shape what can be done with state-owned plantations include history and power structure (frequently including a lack of power to do anything much at all); nature of the plantation asset base; ecological influences and constraints; economic and financial conditions; social–cultural influences and conflicts; institutional norms and precedents; and the scope and rate of change in all of the above.

Option selection and design

How can the 'right' objectives be agreed?

Sharpen up your argument. Actors in the drama – whether they are government policy-makers, company bosses, local groups or NGOs – need to be clear about their own particular reasons for wanting to change ownership and management.

Recognize that other actors have different values. Each set of reasons may represent a different set of scales for weighing costs and benefits.

Encourage transparency amongst actors. Clarity of actors' positions can help to create transparency in the weighing of costs and benefits.

Negotiate practical objectives. Negotiation with key actors is critical in clarifying positions on contested issues, identifying some common objectives, and focusing on practical ways forward which can attract buy-in from those handed new rights and responsibilities for ownership and management.

Arriving at a clear definition of 'the public interest'. The negotiation process should pay particular attention to identifying the public interest in state-owned plantations by asking which goods and services provided by state plantations are threatened by changing ownership and management, and who loses if the goods and services are no longer provided, or if people have to pay for them in future.

Keep objectives clear and simple. Potential opponents are more likely to buy into change if the purpose is clear and they can see beneficial outcomes.

Select the model best suited to purpose. Increased operational efficiency/ profitability can be achieved by outsourcing without incurring the political costs of asset sales and threatened public goods and services. But if you want to encourage small-scale enterprise, far-sighted management and/or concerted investment, major incentives such as transfer of ownership or secure and tradable use rights are essential.

Design the transfer to suit its purpose

Check that the 'design mix' will deliver your objectives. Key drivers behind China's decollectivization have included the desire to raise rural welfare and the need to improve forest land efficiency. To achieve the former, lands have often been allocated on a per capita basis. This approach has not allowed for discrepancies in household productivity and provides little incentive for more efficient producers to specialize in timber production. Moreover, the small size and scattered nature of forest plots is frequently uneconomic. The introduction of forest rights transfers, promotion of shareholding co-operatives and re-allocation of lands that were not being used by their new owners have helped to address this problem (Lu Wenming et al, 2002).

Develop criteria. Develop and promote amongst other actors a set of criteria, based on fundamental elements of sustainable development and in line with national societal priorities, by which decisions about ownership and management transfers can be judged.

Do your homework. Analyse the existing information base and carry out research to examine the options that may deliver the benefits that you want. Ensure that this analysis is guided by sustainable development criteria and identify the powers, rights and responsibilities that need to be transferred and those that need to be retained by government to achieve your aims.

Plan for an optimum balance of powers in line with sustainable development. Aim to transfer all of the rights that private-sector actors need to achieve optimum sustainability objectives and to ensure that government retains the rights necessary to achieve public policy objectives. Enter the negotiation process by clearly making your case for this.

Be prepared for trade-offs. Several aims may sit together but are likely to need reconciliation and compromise (e.g. attracting large-scale investment and encouraging small enterprise development).

Recognize the transaction costs involved. Transaction costs (e.g. in terms of the time required of officials in key ministries) are high even if, in the case of sales to the private sector, there is a realizable sale price. In South Africa, the costs of plantation privatization, when compared against the much greater financial value of other larger privatizations occurring at the same time, appeared disproportionately high and were thus competing for the time and attention of busy officials. In such circumstances there is a danger of a politically easy route to forest privatization being adopted at the expense of longer-term and more difficult issues being resolved. However, in the South Africa case, government did realize that forestry privatization's success should be measured not solely on price and acknowledged its relative importance when linked to land reform and social objectives.

Make it attractive and accessible to your target groups

Make sure the resource is in good condition and free from fundamental conflict. To be of interest to investors and/or communities, resource quality and potential will be a critical determinant, as will the existence of challenges to land use for forestry (e.g. in South Africa, several plantation areas originally offered for sale under lease have now been withdrawn and are in the process of conversion to other land uses).

Ensure transparency of process. Private-sector investors, community groups and, indeed, other government departments often nurture mistrust of government based on past experience of opaque decision-making processes. Making a transfer process attractive to these actors will require clear signals about who will do what and how they will be held accountable.

Build in sufficient security over use rights to encourage investment. Such security is likely to be a function of provisions in a lease or ownership agreement, including duration; the right to assign, sublet and mortgage use rights; the support of the transfer by broader enabling policy and support services derived from, for example, land and institutional reform; and clarity over the ultimate ownership of the land/resource in the case of defined-period leasing.

Contract over a long enough period for the security and planning horizons of contractors and tenants. For example, in South Africa it was agreed following negotiation that the lease is of indefinite duration, but providing for 35 years' notice of termination by either party at any point after the lease's 35th anniversary. This effectively provides a guaranteed minimum tenure of 70 years on entering the lease, with provision for early termination in the event of a material and un-remedied breach of lease conditions. In this example, a key issue is the lessee's confidence in the government's ability to deal with breaches of the lease's terms or the law.

Allow contract transfers. Making the lease assignable/transferable (in whole or in part) to another party makes use rights tradable. An assignable lease has a financial value best protected by practising sound management of the forest. Risks that use rights may be assigned to another, perhaps a non-target, group to realize a quick profit need to be mitigated by requiring government's prior approval of the transfer.

Ensure balanced benefit-sharing. Private actors will not invest if they do not get a sufficient share of the benefits that accrue from their labour and capital. Households who were offered land use rights in China were required to produce a given quota for the collective at a set price or fee. Any excess production could be sold on the free market and this was the main incentive for increased productivity. Since the allocated land tended to be barren and required significant up-front investment, the deals on offer were often unattractive (Lu Wenming et al, 2002).

Balance risks and rewards. The allocation of forest use rights involves the transfer of management and harvesting risks. Risks are particularly high for forestry due to the long time period involved in production. During the early years of China's transfers of collectives, the terms failed to reflect the new allocation of risk and were often heavily weighted against private operators. The small size of plots, short period of tenure and poor quality of land undermined incentives for households to take on risks associated with forest production (Lu Wenming et al, 2002).

Package services, assets or use rights in a way that will attract the target groups. This is best shaped through dialogue between the actors. For example, the South African government carried out various forms of valuation and business analysis to shape the packages of assets offered for privatization; yet, these were contested and modified later when negotiations on the transaction began. Earlier 'market research' consultation with potential buyers might have made the process more efficient.

Ensure clarity of rights and obligations of all parties. Where rights and benefits of households have not been well defined, farmers have often been unconvinced that allocation programmes will last and hesitate to invest capital and labour (Phoung, 2000). Transfer process may need to address insecure tenure associated with unsettled boundary disputes. Under China's system of collective forest management, tenure disputes were rare. With the introduction and extension of contracting out, disputes have re-emerged as people seek control over higher quality land. Such disputes are blamed for causing deforestation in several areas (Lu Wenming et al, 2002).

Address unfavourable investment climates. Even where transfers are made attractive in every other respect, they may still fail if they are set within an unfavourable investment climate. Problems can stem from high taxation, remoteness to markets, expensive finance, the burden of regulation and government bureaucracy and adverse labour relations or costs.

Make sure that your target groups are ready

Estimate capacity. Make an assessment of the capacity of private enterprises to engage profitably with the transaction process and to meet their obligations, as well as make full use of any rights that are transferred.

Promote continuous improvement of management systems. Encourage enterprises or other organizations with the potential to take up new responsibilities for plantations in order to make and implement active internal plans for developing, upgrading and continuously improving their systems for information generation and management, human resource development, participation, planning and management, finance management, and monitoring

Support preparedness in community organizations. In the case of community organizations which are to be granted rights under programmes of transfer of state-owned plantations, their capabilities may appear to be of no concern to the granting body. However, lack of preparation by community organizations may lead to social division and/or destruction and fragmentation of the plantation resource. Community organizations should consider some of the following challenges (Fisher, 2000; Marghescu, 2001; Mayers and Vermeulen, 2002b):

- Generate a high degree of trust among the actors, who must be confident that others will comply with agreements made.
- Building upon existing forms of community organization is usually a better basis for collective action than artificially constructed or administratively convenient units (because people are less likely to breach agreements when doing so will interfere with existing social arrangements).
- Avoid fragmentation with a large number of owners not bound by an umbrella organization or association.
- Ensure complementarity of plantation and social units – collective action for resource management is more likely to occur when the boundaries of the resource and the boundaries of the social unit managing the resource coincide.
- Ensure adequate financing of community management activities.
- Generate sufficient knowledge and expertise about plantation management.
- Overcome conflict within and between community groups.
- Manage the long timeframes involved in tree-growing (separating the benefits from the costs) and the occasional disincentives of seasonality clashes between farming and forestry activities.

Box 3.4 offers some attributes of plantations and community organizations which make them amenable to successful programmes of plantation transfer.

Getting the best deal

Generating and committing to some principles for optimal deals. Whether transfers are planned to private enterprises or community organizations, commitment to some principles of good deal-making will help to produce an effective and equitable result. One possible set of principles is presented in Box 3.5.

Competitive bidding. In the case of transfers to private enterprises, an open-market bidding-based approach to the transaction is crucial. Such auctions can allocate forest land-use rights to the most efficient producer and can maximize sale price and revenue for the contracting authority. However, it is important to recognize that maximization of revenue comes at the expense of public-policy objectives because they are a cost to bidders. Thus, sale price and revenue maximization may not be the most important objective.

Spreading the word. It is vital that clear information about the resource and the proposed transaction process is developed and presented in ways that the target groups can access and digest.

Box 3.4 Attributes of plantation resources and communities that will favour successful transfer of state-owned plantations to community-based organizations: A checklist

Plantation resource attributes (natural capital) include:

- clear and defensible boundaries;
- manageable scale – in spatial terms;
- relative scarcity and/or substantial value;
- relative proximity to communities;
- predictability and ease of monitoring;
- seasonality in tune with livelihoods;
- ease of utilization.

'Community' attributes (social, human, financial and physical capital) include:

- ability to claim and secure tenure;
- small-scale – in social terms;
- demand for, and dependence upon, plantation assets;
- stakeholder identification and group demarcation;
- community/group institutions built on existing motivation (not administrative convenience);
- representation and legitimacy;
- adaptability and resilience;
- effective rules, mutual obligations and sanctions;
- negotiated goals;
- conflict-resolution capability;
- equity in distribution of benefits;
- ability to negotiate with neighbours;
- political space to build community–government relationships;
- confidence to coordinate local operations of external institutions;
- versatile leadership;
- numeracy and literacy;
- strategy and systems for developing and maintaining finance and infrastructure.

Source: adapted from Roe et al (2000) and the case studies in this book

Clarity on risks. Risks and burdens in the transaction will be dealt with by the private sector simply in terms of the price they are willing to pay. The greater the risks the larger the discounting in the price offered for the asset. If the objectives of the transfer and transaction costs of the process are unclear, the price offered will also fall.

Recognize that requiring private-sector creativity to meet public-policy objectives increases risk. A competitive bidding process does not fully proscribe how investors should manage public policy issues; rather, it invites them to use their initiative in responding to them. But it is important not to overburden the transaction with so many public-policy objectives that it becomes unattractive to investors. Objectives such as revitalizing the plantation resource, investing in processing, and maximizing local ownership and employment may all present significant risks to the private sector and will have to be carefully weighed up.

Design the tender system for optimized objectives. The tender system needs to be designed to enable selection of the bidder whose bid best reflects multiple objectives. This need not be the bid with the highest price. Qualitative criteria, such as commitments to future investment and opportunities for local participation and economic empowerment, need to be combined with quantitative criteria such as the price.

Evaluate bids against agreed criteria and each other. Potential investors are invited to compete against each other in response to the agreed criteria by submitting proposals, which might typically include a business plan and an offer price. These are then evaluated against the agreed objectives and each other to identify a preferred investor.

Process and change management

Political buy-in

Employ expertise in policy research, institutional development and good governance. Knowledge and expertise, objectively and impartially given, can help clear the vision of partially sighted supporters and potential opponents of change and lay the opportunities and challenges of change open so that they can be negotiated.

Strengthen relations between decision developers and the ultimate decision-takers. Governments and forest authorities are often unwilling to transfer ownership and management. Forest authorities often do not trust communities, for example, to make the right decisions (Fisher, 2000). A transfer of decision-making authority to local users implies a corresponding reduction in power of the forestry department, which may be resisted by forestry bureaucrats (Khare et al, 2000). Because of this, particularly in 'major privatizations', political interventions are required to break deadlocks; these require identification and careful relationship-building with political and civil servant champions.

Work with the media. The media will work for you to get the positive messages of change across if they are convinced of the arguments, but against you if they sense that the case is weak or that there may be ulterior or hidden motives. Be selective; use the medium that is most effective at reaching your target groups and the newspapers and radio and television channels that follow policies of objective broadcasting, rather than those that might simply be looking for a story which they will spin for maximum impact.

Box 3.5 Possible principles for good plantation transfer deals

- *Mutual respect* of each protagonist's legitimate aims.
- *Fair negotiation* process where protagonists can engage and make informed, transparent and free decisions.
- *Learning approach*: allowing room for disagreement and experimentation.
- *Realistic prospects* of benefits: requires work to accurately predict and secure the mix and allocation of short-, medium- and long-term benefits.
- *Long-term commitment* to transfers as sustainable development ventures (e.g. overcoming short-term risk aversion caused by rises and falls in pulp markets) since both trees and trust take a long time to develop.
- *Risk management*: accurate calculation, allocation and management of risks in terms of public policy, forest production, price and market fluctuation, and in social and environmental terms.
- *Sound business systems*: practical business development principles at the core, not exploitative relationships.
- *Sound livelihoods analysis*: relationships are focused on increasing capital assets of the poor, securing local rights and responsibilities, developing the capacities and comparative advantage of local institutions, and incorporating flexible and dynamic implementation paths.
- *Contribution to broader development strategies* and programmes of community empowerment, and integration or 'nesting' of plantation transfers within wider national and local land use and development frameworks.
- *Independent scrutiny and evaluation* of plantation transfer proposals and monitoring of progress.

Source: adapted from Mayers and Vermeulen (2002a) and the case studies in this book

Buy-in by civil society and private business

Clarify, utilize and build on the complementary skills of the private sector and civil society. The private sector may be particularly strong on management and technical skills; equipment; dissemination and distribution capacity; contacts and sphere of influence; innovation; financial resources/rigour; and fact action. Civil society may be a rich source of other resources, such as on-the-ground know-how; development experience and knowledge; people skills; imaginative and low-cost responses to challenges; social mobilization and public advocacy skills; and associated credibility.

Use issues-based interactions. Bring people to the negotiating table with focused discussions on key issues, not broad, vague agendas. Groundwork can then be done effectively to accommodate all relevant stakeholders, secure buy-in and identify possible risks up front.

Deal with unrealistic expectations. Ensure that key staff have a realistic picture of the scope of the transfer. Maximize regular face-to-face contact with key protagonists. Ensure continuity of approaches and stability of staff in key positions. Try to solve problems while they are still small.

Capacity to manage the process effectively

Get access to technical expertise in a number of core areas, such as forestry management, wood processing, finance, legal and project management, and process management expertise.

Secure a dedicated budget against a reasonably flexible timetable. The costs of plantation transfers need to be carefully planned for, and will include the cost of communicating and providing access to information, raising expectations, specialist skills, a wide range of transactions and stakeholder involvement. Timeframes are also critical; transfer processes require time and flexibility to build trust and momentum, but need to be short enough to prevent information becoming out of date and energy dissipated.

Ensuring that new owners/managers continue to serve the public interest

Conditions of the transaction

The transaction offers opportunities to safeguard against many of the social and environmental losses that people fear may occur following privatization and to deliver specific objectives. Restrictions built into transfers are never free; the beneficiary of the transfer will pass the cost back to the government in the form of a lower purchase price or rental payment.

Specifying minimum equity stakes can ensure that target groups are brought in and safeguard against transfer to a single interest that may not serve all of the government's objectives as effectively as a joint venture.

Restricting rights to the land can be an effective way of protecting the interests of claimants to underlying land rights, while transferring secure ongoing use rights to the new plantation owners, as occurred in South Africa and New Zealand. This can both ensure a good price for the government and a continuing commitment to responsible management.

Obligations written into lease agreements can protect customary rights; but they may be difficult to enforce when the freehold is transferred unless they are protected by law. In contexts where regulation does not enforce good management practice, performance indicators and management standards backed by government or independent audit can protect against damaging environmental and social impacts. Obligations to disclose the results of audits – for example, in the form of annual social and environmental responsibility reports – help to ensure commitment and transparency.

Performance bonds and rights of resumption to the government secure beneficiaries' commitment to complying with the conditions of transfer agreements.

Instruments external to the transaction

Check the forestry governance framework for gaps. The governance framework (see 'Governance of sustainable plantation development and management' in Chapter 2) is key to ensuring that private actors do the things that are required and expected of them. Facilitate national and local strategic fora (e.g. national forestry programme and local governance fora), and other facilities for stakeholder dialogue and to build consensus around forestry goals and forest practice standards. Check that financial incentives (tax relief, direct payments) are pulling managers in the desired direction.

Provide ongoing support to community-based plantation management:

• Convene multi-stakeholder dialogue at national and state levels to transform stale debates and to keep promoting experimentation at the local level.
• Introduce further democratic processes into forestry departments (e.g. training in facilitated frameworks; coherent human resources development policy; formal leeway for front-line staff for experimentation; new forms of reporting, planning and participatory monitoring; and curriculum development and staff placement with other departments).
• Upgrade policy monitoring and analysis capacity.
• Remove legal hurdles to community rights over forest resources.
• Provide clear policy signals to the forest-based private sector in order to engage with local groups and enterprises and to encourage companies to forge direct links with farmers.
• Develop programmes to help tackle intra-community inequity.

Use the power of the market to drive up forest practice standards. Certification can force improvements in forest management (Garforth and Thornber, 2003). A requirement for certification from export markets, particularly in Europe and North America, has stimulated South African producers to keep management standards high (see Chapter 10). In Chile, forest owners accept and are willing to undergo performance-based certification.

But recognize that certification will not work in all situations. Certification is likely to be most effective where:

• There is strong market demand, even though there may be no 'green premium'.
• It is supported by government, in particular state forest agencies.
• There is consensus on standards (which is often hard won).
• In the case of many small growers, there are mechanisms in place to enable cost-sharing and cooperative marketing.

There is no escaping the need for solid practical regulation. In a context of increased reliance on private-sector actors, it might be thought that the emphasis is likely to be on less regulation and greater use of voluntary standards and market instruments, including payments for environmental and social services

(Mayers et al, 2002).There is no evidence from the case study countries of weaker regulation with more liberal economic philosophies. Rather, regulation appears to be a critical prerequisite for achieving a healthy balance between private-sector investment and public needs:

- *Chile*, the most liberal and the first to privatize its state plantations, requires plantation managers to obtain government approval of a management plan before they carry out activities. There are severe penalties for non-compliance (see Chapter 6).
- In the *UK*, forest owners, including plantation owners in England, Scotland and Wales, are required to apply for a licence for permission to fell more than small quantities of timber.The Forestry Commission has the power to impose conditions with the licence and operates a presumption that all felled stands should be regenerated, except where removal of the plantation may result in environmental gains.
- As *China* has shifted collective forests to private households and household groups, the government's ability to achieve forest management standards directly has been severely weakened. In response, the State Forestry Administration has worked hard to develop new forest management standards centred on a system of harvesting certificates. It is also actively considering a system of mandatory forest management certification (Lu Wenming et al, 2002).
- In *India,* The Forest (Conservation) Act 1980 strengthened the central government's control over forests and regulated change in land use, as well as transfer of ownership. Two new clauses added to the act in 1988 have affected the private sector's involvement in forests and in several types of plantation (e.g. rubber, palms and oil-bearing plants) on state forest lands. According to sub-clause 2(iii), any forest land or portion thereof may not be assigned by way of lease or otherwise to any private person or to any authority, corporation, agency or any other organization not managed or controlled by the government without the prior approval of the central government. Sub-clause 2(iv) prohibits clearing of naturally grown trees on forest land for the purpose of using it for afforestation.

Managing the government's post-transaction responsibilities

Powers and duties of 'regulator/supervisory body' need to be clearly defined, appropriate to purpose, understood inside and outside the body and accepted by users. Powers should not be so great as to allow the supervisory body to prevent private actors from making the fullest possible use of their new rights, but should be sufficient to ensure that rights are exercised responsibly.

Motivation and capacity in the supervisory body. Ensure that the culture of the supervisory body supports increased private-sector participation. Prepare staff for their new 'enabling' role through organizational change and training programmes.

4

Conclusions and Ways Forward

Mike Garforth, Natasha Landell-Mills and James Mayers

In this final chapter, before the case studies are presented, we draw conclusions on the extent to which the benefits that governments and other players seek from changing ownership and management of state plantations have been achieved in the countries studied, drawing also from wider experience. We identify the stumbling blocks to successful change processes and suggest how this synthesis of lessons from experience and guidance can be applied in future transfers. Finally, we suggest what needs to be done in order to ensure that initiatives to change ownership and management of state-owned plantations keep improving.

Changing ownership and management: What are the benefits?

Efficiency and profitability of enterprise

Profitability can rise – with some kick-start investment. Privatization of management activities and operations by outsourcing services can reduce costs and increase profitability of state plantation enterprises, in some cases substantially. Gains are achieved when service providers have the capacity and technical competence to deliver the required outputs and when there is sufficient competition. State enterprises need to be prepared to incur costs in order to achieve these two conditions. Costs may take different forms: cash payments to support start-up or development of contracting and training organizations; transfers of machinery; and guarantees of work to take new contracting business through the start-up period.

Outsourcing is efficient, particularly for operations and support services. With few exceptions, outsourcing has focused on plantation operations and supply of materials and support services (e.g. establishment, felling and extraction, seedling production, vehicle and machinery supply, and maintenance), the capacity and skills for which can be developed quickly. Outsourcing of management planning activities may also yield gains, but the evidence is weaker. The private sector certainly can develop the skills for such tasks as providing inventory, environmental assessment and ecological landscape design; the factors that drive the costs of private-sector plantation operations below in-house provision are just as likely to work here.

But there are exceptions. The design of the forest restitution process in European Union (EU) accession countries, geared as it has been to returning forest assets to their former owners or their successors, will probably not yield any efficiency gains (Indufor and Eco, 2001). The small size of the holdings, the lack of interest or capacity, and the skills to engage in forest management all serve to reduce efficiency and profitability.

Investment and innovation

Unleashed creativity brings substantial economic returns. Entrepreneurial drive, encouraged by the transfer of some or all of the assets represented by plantations, has increased the financial return on the assets. Clarke (2000) concludes that the New Zealand privatization has been a positive influence, though it cannot claim all of the credit. In China, small township and village enterprises (TVEs) have developed, many associated with plantation management. TVEs have become the fastest growing component of China's economy. Ninety per cent of China's paper mills are now TVEs, and they have grown more rapidly than the state-owned paper mills (Xu, 1999).

Transaction processes can develop new levels of trust and strategic thinking. South African saw mills are becoming more competitive and profitable after several years of downbeat performance (see Chapter 10). By moving away from a conflict-ridden set of relationships between government and private saw-millers to an integrated one where trees are grown to meet the specifications of improved technology, companies are really starting to think bigger scale and global. If these expectations are achieved, then the process will have laid a foundation that could have a significant impact on the South African economy and contribute local economic development in the areas where forests occur. The homeland forests of South Africa could easily increase productivity by 50 per cent, feeding increased processing capacity and job opportunities.

Privatization can stimulate investment in processing. Chile has a thriving wood processing industry based entirely on privately owned plantations. Clarke (2000) concludes that privatization has facilitated onshore processing in New Zealand and that fears that privatization would encourage log exports have proved unfounded. The South African government's 2003 budget speech had this to say about the privatization process:

> *The restructuring is already bearing fruit. We were delighted to learn that Mondi South Africa is investing 2 billion rand to expand its mill in Richards Bay by some 40 per cent. The project will export pulp worth an additional 500*

million rand per year. Approximately 800 million rand of the investment will be spent in South Africa, with obvious spin-offs for local businesses and their employees, and Mondi has undertaken to use the project to expand procurement from black-owned businesses and its training of historically disadvantaged South Africans. Madam speaker, much of the timber needed will come from Siyaqhubekha Forests: the empowerment company that successfully bid to take over management of the forests around Richards Bay. We should celebrate these developments that the restructuring has unlocked, as well as Mondi's commitment to investment in South Africa.

But conditions have to be just right. Investment in wood processing is not necessarily linked to forest ownership. There is an increasing trend for large wood processors to divest their plantation assets; so the argument that processors need security of wood supply by owning their plantations (e.g. in New Zealand during the 1980s as part of the justification for privatization) does not appear to be holding up. Recent divestments include the 2001 sale by Paperlinx of their plantation estate to Hancock Victorian Plantations Pty Ltd (HVP) for Aus\$152 million in Australia and Fletcher Challenge in New Zealand announcing their intention to divest themselves of their plantation assets. However, there are other examples of processors expanding their plantation assets (e.g. Gunns in Australia). In general, there is no clear argument that owning plantations increases security of wood supply and hence encourages investment in wood processing by the companies owning the plantations. The New Zealand case study suggests that privatization has been successful, to some extent, in encouraging investment; but many believe that tying up capital in purchasing trees limited the capacity to fund investment in downstream processing (see Chapter 9). Other options (i.e. corporatization) may enable more rapid investment in downstream processing than outright sale of plantation assets. Large-scale privatization did not suit the UK's wood processors, who were concerned about losing the security of wood supply tied up in long-term agreements with the Forestry Commission, or non-governmental organizations (NGOs) concerned about the impact on access and other plantation public services. The idea that the Forestry Commission was managing its estate for the nation was too deeply rooted for the government to be able to force through privatization of more than a small proportion of the commission's forests.

Public spending

Outsourcing is an effective way of reducing public spending. Efficiency gains from outsourcing lead directly to reductions in public expenditure. Savings come mainly from reductions in the labour costs, private enterprise having higher levels of productivity per employee.

Gains from asset transfers can be achieved is some circumstances. Net reductions in public spending from asset transfers are most likely to be achieved where the objectives of the state enterprise are focused on exploiting the asset for profit and demand for public goods and services is low, and where the new owners are geared up for management. In China during the 1990s, seven million workers moved out of the state sector after the government permitted the sale of some state-owned enterprises and subsequently permitted managers to release redundant labour (Hyde et al, 2002b). South Africa looks very likely to achieve its objective of unburdening itself from the financial liability of loss-making operations (see Chapter 10).

But asset transfers are not always an effective mechanism. EU accession countries, on the other hand, are faced with serious difficulties in dealing with the additional economic problems resulting from forest restitution (Indufor and Eco, 2001). The costs of addressing lack of capacity and skills in the private sector and the additional regulatory burden are countering any reductions in public spending. In the UK, the additional public spending required to deliver the public goods and services demanded from forests has put any move towards large-scale privatization of state plantations on indefinite hold.

Trade balance

Assets transfers can stimulate exports and reduce imports. Countries with comparative advantage in wood production and manufacture of forest products and a confident private sector can increase domestic production of higher value products for domestic and export markets. Chile has seen steady growth in the volume and value of forest exports since all state plantations were transferred to the private sector during the 1980s. The share of forest products in export value also increased steadily, reaching a high of almost 15 per cent in 1995.

Poverty and livelihoods

Asset transfers can increase income to rural households and communities. In India, Joint Forest Management (JFM) has increased the income and livelihood opportunities for many participating communities. They have benefited from employment generated under JFM projects through, among other things, micro-planning, sales of non-timber forest products (NTFPs) and share in the final harvest; for example, 21.58 million person days of employment were generated in just six states during 2000–2001 (GoI 2002b). In just four states, the forest protection committees (FPCs) received around 65.59 million rupees through benefit-sharing mechanisms during 2000–2001 (GoI 2002b). In West Bengal, though the sharing percentage is the lowest in the country (25 per cent), each FPC is estimated to have received around 70,000 rupees as its share of timber revenue (Palit, 2001). Furthermore, JFM has helped many FPCs to build up substantial levels of community funds, which are used for local development activities. At the end of 2000–2001, total community funds under JFM were 557.09 million rupees in seven states (GoI, 2002b).

But the contribution is not always clear. In China, rural incomes grew more then sixfold in 20 years; but agricultural reforms and the development of smaller TVEs were the primary sources. Forest income was a supplement to greater household income from other sources (see Chapter 7).

And benefits may not be distributed fairly. Households at all levels may benefit; but better-off households are likely to obtain the greatest proportional benefits (Ruiz-Perez et al, 2002).

Transfers can stimulate higher levels of sustainable production. In Harda Division of Madhya Pradesh, irrigation facilities developed under JFM have increased the crop yield by two to five times. In Gujarat, better availability of grass and tree fodder after the initiation of JFM has led to an increase in milk production in several villages. For example, in Nisana village (Vyara Division), it has gone up from 40,000 to 200,000 litres per year. In some states, FPCs have started earning through sale of produce from their forest patches. In four states, FPCs received 62.59 million rupees through benefit-sharing under JFM during 2001–2002. Income from NTFP is generally more than the share in timber revenue (see Chapter 8).

Disadvantaged groups can benefit. Women in several FPCs in West Bengal are able to earn between 4500 and 6000 rupees annually through the sale of *sal* leaf plates (see Chapter 8). Transfers to companies with black shareholders and to rural black communities will contribute to addressing the injustices of the past (see Chapter 10).

Governance

Transfers can change attitudes and relationships for the better. One of the most significant impacts of the JFM programme has been the change in attitudes of local communities and forest officials towards each other and forests. For instance, members of Botha FPC in Buldhana, Maharashtra, even postponed a wedding in the village in order to fight a forest fire. This was unthinkable in the pre-JFM days. In several FPCs, traditional forest protection practices have been revived – for example, *kesar chhanta* (sacred groves) in Rajasthan (see Chapter 8). There is also greater acceptance of participatory approaches among forest protection officials. The large number of training and orientation exercises carried out in different states has also contributed to the positive change in attitude. The magnitude of the effort can be gauged from just one state, Andhra Pradesh, where more than 20,000 JFM-related training programmes have been carried out in recent years (Mukherjee, 2001). In South Africa, privatization deals require private-sector managers to create good relationships with surrounding communities. Not only is this an essential criteria which is assessed for certification purposes, but it is regarded as 'just good practice' in many areas. The risk of fire is too great for managers not to invest in working with the communities to build trust. Providing access to the forests for household goods and services will contribute towards this relationship-building.

And enhance social capital. The institutional capacity created to manage income streams to new community owners in South Africa will provide a useful platform upon which other local economic development initiatives can be managed (see Chapter 10). The UK's limited experience of transferring ownership and management to rural communities points to greater community cohesion and inclination and the capacity to act cooperatively (Haggith, personal communication, citing the examples of Laggan and Culag Community Woodland near Lochinver).

Transfers can stimulate greater involvement of other stakeholders. While the JFM programme has created greater space for community participation in forest management, it has also led to greater involvement of other stakeholders, such as NGOs and *panchayats*, and helped to bring greater transparency to the sector. Information from six states reveals that over 1000 NGOs are actively participating in JFM. In some states such as Madhya Pradesh and Uttar Pradesh, Forest Department officials and NGOs are working together at the field level in the form of *Spearhead Teams* (GoI 2002b).

And promote leadership and awareness. The discussions between Forest Department officials and NGO staff members, as well as exposure visits and training programmes, have catalysed the development of leadership in villages. Better information flow has made people more aware about various government policies, as well as their rights. This has had a positive impact beyond the forest sector.

Ecological processes and environmental health

Engaging communities as partners helps to combat forest degradation and improve forest condition. There is evidence that JFM has improved the condition of India's forests. During the past few years, the overall forest cover has increase by 3896 square kilometres and dense forest cover by 10,098 square kilometres, much of it attributed to the success of JFM (FSI, 1999). In areas under JFM, incidents of illegal felling have declined sharply. It has been reported that in Rajasthan, unlike during the past, people did not resort to felling in JFM areas even during droughts (Ghose, 2001). JFM has helped to reduce the area under illegal encroachment as well as the rate of fresh encroachments. For example, in Andhra Pradesh, nearly 12 per cent of the encroached forest land (38,158 hectares) has reportedly been vacated since the JFM programme was initiated (Mukherjee, 2001).

And standards can rise under the management of private corporations. South Africa's lease agreements require private-sector managers to have their forests certified in terms of an internationally recognized certification agency. Many of the former homeland forests had become severely degraded and were deteriorating up until the point of transfer. Since transfer, these forests have been integrated within the certification programmes of the private companies and significant improvements have been made. Streams have been cleared, invasive plants removed, and catchment management has generally improved.

Stumbling blocks

Some outcomes are hard won

The experience of countries that have entered the arena of changing ownership and management has been generally positive. But some outcomes are harder won than others. Efficiency, profitability and public spending gains from outsourcing are the most transparently dramatic gains that can be attributed entirely to change. Net reductions in public spending and national welfare gains from transfer of use rights or ownership are more difficult to demonstrate if the transaction costs and the additional costs of monitoring and enforcement are taken into account.

Most difficult to achieve have been the livelihood outcomes sought from transfers to rural communities. Problems of community technical and financial capacity and human capital, lack of clarity and security over tenure, and the barriers erected by government institutions continue to work against success. All of these problems can be overcome given imagination and determination. The experience of Joint Forest Management in India, of village and household entrepreneurship in China and the expectations of successful transfers to rural South African communities are cause for optimism. Countries such as the UK that have had less success could learn a lot from their experience.

Other outcomes are not wanted

We should expect unwanted outcomes, some of which we can anticipate and attempt to avoid or mitigate, and others that we had not expected. The evidence is that change can be effected without significant negative effects provided that the potential impacts of change are assessed and avoidance or mitigation measures taken within the transfer process, or in the broader governance framework within which transfers take place.

Impact on jobs. Efficiency gains invariably mean job losses; but they can usually be anticipated and mitigation measures taken, as in South Africa. Clarke (2000) concludes that in New Zealand the impact is unclear; but given the low labour intensity of forestry in New Zealand, it is more of a perceived than a real issue. The trend of increasing mechanization in plantation operations in the residual state sector will force job losses, anyway. Changing ownership and management may offer options for reducing labour forces that are more equitable than occasional bouts of job shedding.

Loss of forest cover linked to restitution has not occurred in most European Community Accession (ECA) countries (Indufor and Eco, 2001). South Africa, on the other hand, expects a reduction of forest area in the short to medium term as government pulls out of non-viable operations. The future of the Department of Water Affairs and Forestry's (DWAF's) category B plantations is of particular concern, though this is less to do with loss of plantation area and more to do with job losses.

Quality of management may deteriorate where new owners and managers do not have the capacity or sufficient financial incentive. Indufor and Eco (2001) observe that neglect of forest management and a general backlog of necessary tending operations in private forests may turn out to be more of a problem than over-harvesting. This is an inevitable consequence of the policy of returning expropriated forests to private owners with no capacity or motivation in small lots. The problem could have been addressed by preparing the new owners for their rights and responsibilities and holding back from restitution until the conditions for success were met.

Ways forward

How to use this synthesis of lessons from experience and guidance

Actors in existing transfer processes and potential new ones should enter the arena with their eyes open, with this book as a source of ideas, and with commitment to help the process meet some basic requirements:

- appropriate stakeholder engagement methods;
- agreed principles for the transfer;
- a proper understanding of all stakeholders;
- catalysts for the process;
- specific activities and steps;
- a phased and 'learning' approach – with room to experiment, fail, succeed and adapt;
- adequate resources, skills and time;
- demonstrable results and benefits, especially some 'early wins' to bring people on board and build momentum.

Policy-makers need to clearly define the goals of privatization and design the form of privatization to meet these goals. Different levels and modes of privatization will meet different goals. For example, corporatization may be the best option if wanting to encourage downstream investment in processing rather than tying up capital in trees. Corporatization achieved significant efficiency gains with the New Zealand short-lived experience, demonstrating a rapid gain in economic efficiency. In other words, outright sale is not the only option that can achieve the goals often aimed for in privatization. In addition, corporatization, which improves efficiency and profits, can work usefully to increase the value of the asset for subsequent sale.

Privatization does not happen at a stroke; it may take years for the state to divest itself of its stock of transferable plantations. Momentum needs to be maintained and the process kept under review so that opportunities can be taken to improve on the returns to the state and the beneficiaries and to avoid unwanted outcomes.

How can we continue to learn?

Plantation transfer is not about trees; it is about people and power. Transfers of ownership can be the right thing when they put power in the hands of those who can use plantations for equitable, efficient and sustainable ends. They can be the 'doable' thing when there is absolute clarity of purpose and dedication of practitioners. But transfers can also go astray and be used to concentrate plantation power and privilege in too few hands. Those seeking sustainable solutions need to use all of the tactics that they can muster to spread sound knowledge and shape the political process such that it finds these solutions.

Evaluation methodologies need to be developed and implemented to provide a consistent approach to assessing the success of existing privatization. Assessments need to be against a range of criteria (social, economic and environmental) and include assessment of the adequacy of regulation of plantations post-privatization. Criteria and indicator and certification frameworks may offer the best monitoring evaluation frameworks.

Part 2
Country Case Studies

A Mixed Economy Commonwealth of States: Australia

Jacki Schirmer and Peter Kanowski

Introduction: people and forests in a dry continent

Australia is the driest inhabited continent in the world, with over one third of the country classified as arid (< 250mm average rainfall annually) and another third semi-arid (250–500mm rainfall annually).

Both climate and rainfall are highly variable across much of the country, and soils are generally poor. For these reasons, the majority of Australia's forests, and of its 20 million people and their activities, are concentrated in a band within 200 to 400km of the eastern, south-eastern and south-western coasts of the continent, and in the island state of Tasmania (CoA, 1996, 2002). Most plantation forests[1] have been established within this zone.

Australia's first people arrived approximately 60,000 years ago, and their land management practices – principally their extensive use of fire – altered the distribution, structure and biodiversity of the native forests.

The displacement of indigenous people by European settlers from 1788 also altered the forests directly and profoundly – through their conversion to agriculture, their exploitation for wood and other products, the introduction of exotic animals and plants, and the associated disruption of ecosystem processes.

Australia's natural forests

Australia's natural forests[2] are extensive, diverse and globally unique, but now occupy only 20 per cent: some 156 million hectares of the country's land area, representing about 60 per cent of the area forested at the time of first European settlement (NFI, 1998). The majority, some 112 million hectares (72 per cent), are defined as woodland[3] and have little direct commercial

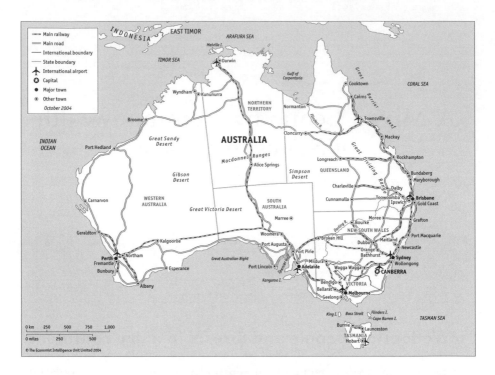

Figure 5.1 *Australia*

role. Despite substantial rates of both commercial and non-commercial reforestation, there is still a net national loss of natural forest annually as a consequence of the conversion of approximately 400,000ha of forest annually to agriculture (CoA, 2002).[4]

Nationally, nearly 30 per cent of natural forests are privately owned and managed. A further 40 per cent are formally under public ownership, but are held and managed under lease by the private sector, principally for grazing, with varying forest rights and responsibilities. The remaining 30 per cent are publicly owned and managed: 11 per cent are in conservation reserves; 9 per cent are managed for multiple use, including wood production; and 10 per cent are held under other tenures (NFI, 1998).[5]

Plantation forests

Commercial plantation forests have been established in Australia since the 1870s. Until the early 1980s, they were established principally with exotic softwoods on sites converted directly from 'less productive' natural forests. Over the past 25 years most plantations have been established on sites previously cleared for agriculture as natural forest sites became increasingly valued for their other attributes. Almost all Australian plantations are rain-fed on sites which receive more than 750mm of rainfall annually, although there has recently been a greater focus on developing plantation forestry options suitable for lower rainfall sites (see, for example, Williams et al, 2001).

Legislation and policy

Commonwealth government

The Commonwealth government has no direct powers over land management in Australia, except on the small area of land under Commonwealth tenure. Despite this, the Commonwealth has exerted considerable influence over natural resource and forest policy in the states in a number of ways (Carron, 1985; Cochrane and Gerritsen, 1990; Dargavel, 1995; NRRPC 2001):

- *Distribution of funds*: this involves the tying of state budget allocations to specific purposes, and making grants to the states for particular activities.
- *Commonwealth–state bodies working to develop national policy*: the Australian Forestry Council has worked to develop common agreement on plantation policy and initiated the Softwood Forestry Agreement Acts in 1967 and 1972. The Commonwealth and states also work together at ministerial and department head level through what is now known as the Forestry and Forest Products Council (FFPC) (CoA, 2002).
- *Joint Commonwealth–state initiatives*: the *Plantations for Australia: the 2020 Vision* (MCFFA, 1997) is the primary example of this.
- *Incentives[6] for the private sector*: incentive mechanisms used to promote forest plantations in Australia have been reviewed by Miller (2002). They include industry and trade policy instruments and taxation regimes.
- *Research and development*: Plantation-sector R&D is funded, in part, through various Commonwealth programmes, such as the Research and Development Corporations (e.g. for forests and wood products; FWPRDC, 2002) and the Cooperative Research Centres (e.g. for sustainable production forestry; CRC–SPF, 2002). The majority of R&D funding is joint, with the forest industries and Commonwealth and state agencies.
- *Export controls*: quotas for timber export (principally for native forest products) were set by the Commonwealth government under the 1982 Export Control Act, implemented in response to concerns about the environmental impacts of native forest harvesting in Australia.

State governments

The state governments have direct powers to legislate for land management, and a variety of legislation affecting plantation management has been passed in each state. Wilkinson (1999) reviewed forest practices systems, generally, and for five Australian states, and Stanton (2000) reviewed those for Australian plantation forests. Most have recently been – or are in the process of being – developed as more contemporary 'new generation' instruments (Gunningham and Sinclair, 2001), consistent with international environmental management system standards, and relying more on self-regulation with independent audit (Wilkinson, 1999). In most states, codes of practice for forest management have been implemented. However, there remain differences in many Australian states between the regulatory treatment of public and private forests, and between that of plantation forestry and traditional agricultural land uses.

The substantial increase in legislation relating to environmental management since the 1960s has led to concerns on the part of many associated with the plantation industry that 'green tape' has become an impediment to plantation forestry expansion. Different states have responded to these concerns in different ways, as the examples below illustrate:

New South Wales
Until 2000 in New South Wales (NSW), private growers had to obtain up to ten or more licences, approvals or permits under the requirements of five different acts before being able to establish a plantation (DLWC, 2001) There was concern that this was an obstacle to private investment in plantations, and that the major state agency establishing plantations, State Forests of New South Wales (SFNSW), was also being adversely affected by it. In 2000, the NSW government passed the Plantations and Reafforestation Act, which aimed to consolidate all of the different requirements of the acts, and made the Department of Land and Water Conservation (DLWC) the sole consent authority for establishing new plantations (DLWC, 2001).

Western Australia
In Western Australia (WA), by contrast, there are still several agencies which may potentially regulate the establishment of new plantations, rather than a single consent authority. However, the requirements of different legislation have been consolidated and interpreted in the Code of Practices for Timber Plantations in Western Australia. The code is a voluntary industry code, which assists companies by setting out how they should comply with relevant legislation and where a permit or approval may be required from particular agencies.

Tasmania
In Tasmania, the forest practices system is the principal regulatory mechanism for forestry on all tenures. It is administered through a single body, the Forest Practices Board (FPB), under the Forest Practices Act 1985. The Forest Practices Code sets out how forest practices are to be conducted in order to provide appropriate protection to the environment and is enforced by forest practices officers (FPB, 2000).

Local government

Local government has planning powers over privately owned land, but no powers over publicly owned land. Therefore, it has played no part in the regulation of plantations established on public land, which form the majority of state-owned plantations. However, it does play a significant role in the regulation of privately owned plantations in some states. One of the objectives of the *Plantations for Australia: the 2020 Vision* has been to ensure that the controls applied on the establishment and management of plantations are reasonable and relatively consistent across different local governments. Some states (e.g. Victoria) require plantation companies to submit plans to local government before establishing plantations; others have encouraged plantations to be considered an 'as of right use' in agricultural zones. In Tasmania, private landholders may apply to have land declared a 'private timber reserve'; thereby removing forest – including plantation forest – management decisions associated with that land from local government control.

Regulatory roles of different governments

In general, regulation of plantation forestry at state level has focused on regulating the environmental impacts of plantations. At the local government level, however, there have been some attempts to implement planning policy relating to the location and type of plantations established on the basis of the perceived potential social and economic impacts of those plantations. This has often proved to be controversial as many rural landholders believe planning guidelines based on economic and social preferences represent an unfair degree of regulation of their choice of land use; similarly, state and Commonwealth policies, such as the *Plantations for Australia: the 2020 Vision,* seek to ensure that there are no differential impediments to establishing plantations, and such planning restrictions are seen in that light. In some recent cases, local government planning decisions that placed restrictions on plantation establishment have been overridden by the state government (Schirmer, 2002a). Similar issues have also recently arisen in some plantation regions in relation to concerns about the water use of forest plantations relative to that of other agricultural land uses. These tensions between plantation forestry and other forms of primary production pose challenges to governments seeking to balance rural community concerns about plantations, with the wider public policy intent of providing a 'level playing field' for plantation establishment.

Australia's plantation industries

Contribution of plantations to the national economy

Australia's forest industries have been estimated to contribute approximately 1 per cent of Australia's gross domestic product (GDP) (NFI, 1998). This does not include the contribution of large users of wood-based forest products, such as the construction and building industries.

In volume terms, Australia is a net exporter of forest products, with a net export of 4.7 million cubic metres of forest products in 2001; but in terms of value, Australia is a net importer, importing a net Aus$1,622.35 million of forest products in 2001. This reflects the low value of a large proportion of the wood products being exported from Australia (ABARE, 2001), the majority of which are eucalypt woodchips.

It is difficult to determine the value of the contribution of plantations to the national economy, as neither the Australian Bureau of Statistics (ABS), nor the Australian Bureau of Agricultural and Resource Economics (ABARE) separate their reporting on forest products industries into plantation and native forest wood sources (Clark, 2002).

Burns et al (1999) estimated that, in 1996–1997, plantation-derived forest products comprised Aus$3.9 billion of the approximately Aus$6 billion of forest products produced by Australia's forest industries, and that around 16,000 people were employed in plantation management and processing, comprising around one quarter of the total number employed in the forest industries.

Clark (2002) has disaggregated Australian production of wood and wood products by wood source. Her estimates indicate that plantation production as a proportion of total production has increased rapidly during the past decade, and now provides the majority of wood used in domestic sawn timber and wood panel production, domestic pulp and paper production, and other domestically processed finished wood products.

Contribution of plantations to regional economies

There are few studies of the contribution of the plantation forestry sector to Australian regional economies. The most comprehensive of the plantation industry itself was that conducted for the Oberon region of NSW, a major softwood plantation growing and processing centre (Dwyer, Leslie and Powell, 1995); more recently, Wareing et al (2002) reviewed the socio-economic importance of the timber industry, based on both native and plantation forests, in North-east Victoria, and Petheram et al (2000) assessed the socio-economic impact of land use change from agriculture to plantation forestry in South-west Victoria.

The conclusions of these studies and reviews are broadly consistent. Investment in growing and processing plantations injected capital and provided employment opportunities and flow-on benefits into regional economies, with most significant employment and regional economic and social benefits generated by investment in value-adding processing. However, as these facilities are increasingly large scale and capital intensive, there are only modest direct employment benefits to be had from their expansion – and these are typically concentrated around the larger regional centres rather than across the region (Stayner, 1999).

The plantation estate

Australia's total plantation estate in December 2003 was estimated to be 1,664,185ha. Of this, 59.4 per cent were softwoods – principally the exotic taxa *Pinus radiata* – and 40.6 per cent were hardwoods – almost all *Eucalyptus* (NFI, 2004). Softwood plantations are grown on rotations of 25 to 45 years for solid wood production, and hardwood plantations on shorter rotations of 10 to 15 years for pulpwood production (Wood et al, 2001).

Figure 5.2 shows Australia's current plantation estate by planting period. Until the 1980s, the plantations established were predominantly softwood plantations. Since the 1980s, and particularly since 1990, hardwood species have increasingly been established.

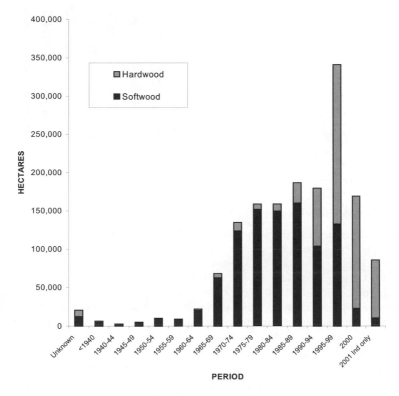

Source: Wood et al (2001); NFI (2002)

1 The figures up to 2000 show the standing plantation estate by planting period as at 2000; the planting period in some cases indicates the most recent planting, rather than the year in which the first rotation of the plantation was established

2 The figures up to 2000 include both industrial and farm forestry plantings; 2001 figures include only industrial plantings

Figure 5.2 *Area of plantations by planting period in Australia*

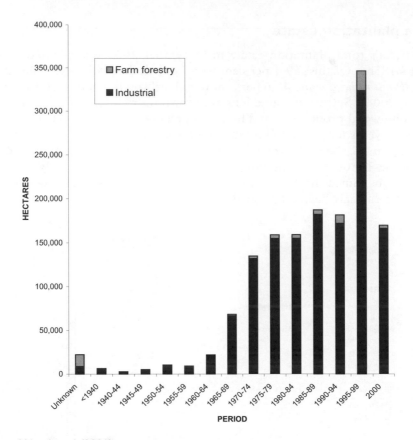

Source: Wood et al (2001)

1 The figures up to 2000 show the standing plantation estate by planting period as at 2000; the planting period in some cases indicates the most recent planting, rather than the year in which the first rotation of the plantation was established

2 Industrial plantations refer to plantations established by growers who manage a combined total estate of > 1000 hectares, including joint ventures where one partner is a large grower and some industrial companies with estates of < 1000 hectares. Farm forestry refers to plantations owned outright by individuals who have a total plantation estate of under 1000 hectares

Figure 5.3 *Area of industrial plantations and farm forestry by planting period in Australia*

Some 95 per cent of Australia's plantation estate is classified as industrial plantations and 5 per cent as various forms of farm forestry (Wood et al, 2001). Most of the farm forestry estate is currently in its first rotation, and the expansion in farm forestry planting since 1980 can be seen in Figure 5.3.

Figure 5.4 shows the area of plantations established by land and tree ownership. Nationally, there has been a significant shift over the past three decades to private ownership of plantations. Fifty-six per cent of the current

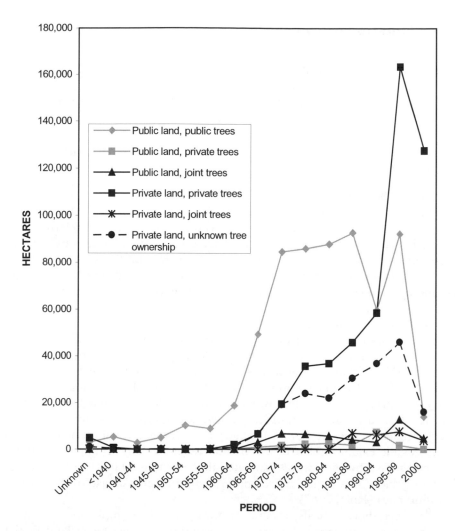

Source: Wood et al (2001)

Figure 5.4 *Area of plantations in Australia by land and tree ownership*

plantation estate is on privately owned land, while approximately 54 per cent of the trees are privately owned, 6 per cent are jointly owned by public and private organizations, and 40 per cent are publicly owned (NFI, 2002). During the 1960s, by contrast, around 75 per cent of plantations were owned by state forestry agencies. There remains considerable variation between the states (NFI, 2002).

The majority of the total area of Australia's plantation estate is owned and/or managed by a relatively small number of organizations. State forest agencies generally remain the largest single owners of forest plantations, despite the increasing trend towards private ownership.

The plantation processing industry

In 2001, there were approximately 270 softwood mills producing sawn timber from plantations in Australia. Additionally, there were 22 mills producing pulp or paper products and 30 veneer and board mills – a large proportion of which processed plantation products (AFFA, 2002). Several woodchip mills also operate in Australia and many of these now also use plantation-sourced logs.

State-owned plantations: Development and evolution, 1880–2002

Initiation: 1880–1920

When Europeans settled Australia, they found a large supply of hardwood timber in the native forests. Native softwoods, however, were relatively restricted geographically. During the 1880s, experimental plantings by government agencies, on land converted directly from natural forest, led gradually to the selection of appropriate species and the development of silvicultural techniques that suited Australian conditions (Carron, 1985).

Consolidation: 1920–1967

The early establishment of plantations, until the 1960s, was on a relatively small scale. Planting programmes were started and then interrupted by World War I and World War II, after which they increased significantly.

In 1964, the Australian Forestry Council was formed as a standing advisory body, comprising the Commonwealth Ministers for National Development and for Territories and each of the ministers responsible for forestry in the six Australian states. At its first formal meeting, the council set as one of its priorities the establishment of a softwood resource large enough to achieve national self-sufficiency; it was estimated that 30,000ha should be planted per year to achieve this target, representing a significant increase over previous planting rates (Rule, 1967).

However, there were also secondary motives, such as employment creation and the support of regional economic development. The Queensland Forest Service, for example, sought to develop plantation resources sufficient to support local-scale industries at geographically dispersed locations along the Queensland coast (Carron, 1985). There were also, in most states at some stage, various small-scale employment schemes based on plantation establishment. In Tasmania, for example, major plantings were undertaken at Fingal and Strahan from the early 1960s to alleviate local unemployment (Inglis et al, 1985).

Expansion: 1967–1980

To help achieve the expanded planting rate set in 1964, the two Softwood Forestry Agreements Acts were passed in 1967 and 1972 (Standing Committee on Environment and Conservation 1975). Under these acts, the Commonwealth government loaned money to the State governments for the establishment of plantations; consequently, establishment rates increased significantly (Wood et al, 2001).

The rapid expansion of plantations coincided with the rise of the modern form of the environmental movement, and the clearing of native forests for the establishment of plantations was criticized strongly (see, for example, Routley and Routley, 1975; Dargavel, 1995). At the same time, the quest for self-sufficiency was being questioned and was discarded in 1974 (Standing Committee on Environment and Conservation, 1975). In 1975, the 1967 and 1972 Softwood Forestry Agreements Acts were reviewed and it was concluded that, in most cases, money should not be loaned for the planting of softwoods in areas where native forest was to be clear-felled; instead, it was argued that marginal agricultural land be used for future plantations (Standing Committee on Environment and Conservation, 1975, pp8, 50).

Mixing the public and private sector: Joint venture/leasing, 1980–2002

The expansion of Australia's plantation estate remained a public-policy priority, with proposals made during the late 1980s by the Forestry and Forest Products Industry Council (FAFPIC, 1987) and the Australian Conservation Foundation (ACF, 1988)

As a result of public opposition, conversion of natural forests to plantations was phased out in most states from the 1980s, now occurring only in Tasmania. Suitable cleared, usually ex-agricultural, land was therefore needed to expand the estates and almost all of this was under private ownership. However, resistance by rural communities against the state purchase of privately owned agricultural land for plantation establishment (Rodger, 1952; SPIS, 1990) and the cost of purchasing such land meant that new ways of achieving the desired expansion of the plantation estate were needed.

Consequently, a shift towards joint private–public plantation establishment began during the late 1980s, accompanying policy goals of returning trees to agricultural landscapes in order to address unsustainable land-use practices and diversify farm incomes (Shea and Bartle, 1988; Williams et al, 2001).

Joint venture and leasing programmes have been the two common forms of partnership arrangement used. Joint venture programmes usually involve private landholders supplying land, while the public agency provides funds for the plantation establishment. The public agency establishes and manages the plantation, and returns from harvest are shared between the joint venture partners at the time of harvest.

Annuity, or leasing, schemes usually involve paying a private landholder an annual fee for the use of their land for growing a plantation. The costs of growing the plantation are met by the public agency in return for 100 per cent, or close to 100 per cent, of returns at final harvest.

Corporatization and privatization: 1990–2002[7]

Corporatization All of the state forest agencies originally operated as government departments which had a range of goals, many of them non-commercial. From the 1980s, commercial objectives became important, and the state forest agencies were restructured to operate as commercial entities rather than as government departments. All of the state forest agencies now operate as government trading enterprises (GTE) or as business units within government departments.

The shift towards corporatization of state forest agencies was part of a nationwide trend, motivated by a range of reasons, including reducing government budget deficits and improving economic efficiency. Since 1995, moves to corporatization have been driven largely by the review of government businesses under the National Competition Policy (NCP), which aims to remove any unfair advantages that government businesses may have had over competitors in the private sector.

The primary changes directly related to corporatization of the different state forest agencies in relation to plantations have been:

• a clearly stated primary goal of focusing on the commercial business of managing the plantation;
• adoption of accounting procedures that are similar to private-sector corporate accounting, with transparent financial statements and audit requirements;
• a separation, in some cases, of regulatory and business functions with a government department taking on the regulation and oversight role, while the new GTE becomes solely responsible for the business activities related to the state-owned plantations.

The effect of corporatization on employment levels is difficult to estimate. In many of the agencies, there was downsizing, resulting both from the outsourcing of work and from the shift of focus from a wide range of objectives to that of a commercial enterprise. However, corporatization does not appear to have adversely affected the provision of various services that are not directly related to the core business of growing, harvesting and selling plantations. For example, agencies that provided extension services prior to corporatization have generally continued to do so since corporatization.

Privatization To date, the only case of full privatization of a state-owned plantation resource in Australia has been the sale of the Victorian Plantation Corporation (ViPC) to Hancock Victorian Plantations Pty Ltd (HVP), a subsidiary of the US-based Hancock Timber Resource Group in 1998. The ViPC was sold through a direct sale rather than as a share float as has been the case with some other privatizations in the country. Under the privatization, HVP has a lease over the land that was vested in the ViPC, along with ownership of the trees and the right to harvest and replant them in perpetuity, other than a small area over which only a short-term lease was granted to allow indigenous species to naturally regenerate after harvest of the plantation (Victorian Auditor-General's Office, 1999). This trade option was chosen because it was anticipated that it would generate the greatest return to the state.

The ViPC privatization was achieved in four stages:

1   the establishment of the ViPC in 1993, under which the plantation business was treated as a sole entity and on a commercial basis rather than as a government department that had multiple facets and competing budgetary obligations;
2   the commercialization of longstanding wood-supply agreements, most of which were enshrined in acts of parliament;
3   surveying of plantation lands vested in the ViPC by accurate delineation;
4   an international offer for sale, which was managed by the Victorian Treasury and international consultants.

Privatisation of ViPC continued the changes begun when it was commercialised. Principle amongst these were:

*   a sharper focus on plantation management as a business enterprise, rather than on a wider range of responsibilities, and on operating as a dedicated, commercial plantation company;
*   an associated reduction of the total workforce employed by the organization;
*   stronger linkages to international markets as the result of purchase by an international investor;
*   a more explicit policy of corporate accountability to key stakeholders, and the development of processes for dialogue with those stakeholders.

Current utilization and management of state-owned plantations

This section refers to plantations owned either solely by the state or jointly by the public and private sectors.

Management of state-owned plantations: Interactions with the private sector

Plantation establishment, management and harvesting

A large proportion of plantation establishment and management activities was, in most states, originally carried out by workers employed by state forest agencies. This has changed over recent decades, with work increasingly being contracted out to private operators, generally through a tender process. For example, Forestry Tasmania has shifted from conducting 70 per cent of its direct operations using its own resources prior to corporatization, to contracting out around 70 per cent of its direct operations to the private sector.

The state plantation agencies generally use their own staff to plan, manage and audit plantation operations. Harvesting work has traditionally been carried out by the private sector and this has not changed with corporatization.

Plantation wood processing and pricing

The plantation processing sector in Australia has always been predominantly privately owned, with the exception of South Australia (SA), where the state Woods and Forests Department had its own plantation softwood processing facilities. Historically, the principle objective of state forest plantation growers has been to supply private-sector processors with wood suitable for value-adding, commonly under long-term supply agreements. However, there have been concerns that these agreements involve setting administratively determined prices for timber which do not necessarily take into account the actual costs involved in growing it (Byron and Douglas, 1981; Hurley, 1986; Cochrane and Gerritsen, 1990).

Commercialization and corporatization of state-owned plantation operations has been accompanied by changes in the methods of determining prices for timber, with cost-based procedures increasingly implemented. The National Competition Policy (NCP) encourages forest agencies to 'ensure that the goods and services they supply are priced to cover their full costs of production, including ... costs ... to which private businesses are normally subject' (Trembath, 2002). A range of methods has been proposed for estimating market equivalent prices, including using competitive auctions and residual valuation techniques. However, as there are relatively few growers and processors in the industry, a competitive auction of timber may not achieve a price reflecting true market value. Recommendations that log residual values be used to set prices have also been criticized, with concern that this may lead to prices being determined by the 'ability to pay' of the processing industry, rather than by the cost of growing the wood (Trembath, 2002).

User rights in state-owned plantations

Recreation

The predominant non-timber use of plantations is for recreation. The overall volume of recreational use of plantation areas is relatively small in most states except SA and the Australian Capital Territory (ACT). In South Australia, the relatively small area of native forests, and in the ACT the proximity of plantations to urban areas, have led to heavy recreational usage and the development of facilities, including walking trails and amenities in plantation areas.

In general, public plantation areas are open to public access, although state forestry agencies have the right to restrict access in certain circumstances – for example, during harvesting operations and when there is high fire risk. Plantations established by public agencies on private land, however, are not generally open for public recreation. These types of plantations constitute a large proportion of the new plantations being established, suggesting that the area of plantations available for recreation is not likely to increase in the future.

Grazing

Grazing on a fee basis is allowed at some stages of plantation growth in some publicly owned plantations; although it is generally a minor part of plantation business, grazing rights are often keenly sought and retained by neighbours or other graziers. Grazing in plantation areas is likely to occur on a larger scale in joint venture and lease plantations, where it would be undertaken by the private landholder whose land is contracted to the plantation grower.

Firewood

Firewood collection is also allowed by some state forest agencies. In the ACT and SA, firewood collectors are allowed to gather firewood in plantation areas after purchasing a permit, and are only allowed to take wood lying on the ground. In West Australia, firewood collectors may collect wood residues from the forest floor after thinning or clear-felling. HVP does not allow commercial firewood collection in its plantations.

Private plantations

This section refers to plantations that are predominantly privately owned. Plantations owned jointly between the public and private sector were discussed in the previous section. Three forms of private plantation expansion have occurred, associated with investments by wood product companies, by investment companies and by small-scale individual growers.

Plantations established by wood product companies

Some plantations have been established by subsidiaries of companies involved in wood processing. These plantations have usually been established as a way of developing a resource owned by the company which allows the company to control and, hence, achieve certainty of supply of a part of their wood supply for future processing (McKenzie Smith, 1975). In some cases, these companies received assistance from the state to establish plantations – for example, through forms of investment assistance.

Historically, establishment of plantations by industrial companies has attracted relatively little criticism or controversy compared to that associated with private plantations established by investment companies (McKenzie Smith, 1975). More recently, public opinion does not appear to have differentiated between the two forms of ownership.

From the 1980s onwards, some industrial companies began establishing joint venture and leasing schemes, similar to those used by the public sector described above, under which they established plantations on agricultural land with private landholders. The proportion of this type of plantation establishment is unknown.

Plantation established by investment companies

During the 1920s and 1930s, several companies began selling 'bonds' in pre-dominantly softwood plantations to investors and promising to establish and manage plantations and provide a return at final harvest. The majority of these schemes failed during the 1930s and 1940s. It was perhaps at this time that investment schemes in plantations began to achieve a notorious reputation, as some investors discovered that their money had either not been invested in plantations, or had been invested in plantations that did not grow or achieve returns as predicted at the time of investment (McKenzie Smith, 1975). When bond companies and new forms of investment prospectus companies began to invest in plantations again in the 1960s, a similar experience occurred (McKenzie Smith, 1975).

The contemporary form of investment prospectus companies began to be established during the late 1980s and the 1990s. They operated according to much stricter rules regarding investment structures than earlier companies. These investment companies were established in response to a variety of events, including the technical capacity to grow short-rotation hardwood eucalyptus crops for woodchip export; strong state and Commonwealth support for the expansion of plantations on privately owned land; and the presence of tax deductibility provisions that provide an incentive for investors to put their money into an investment that has a relatively long time before return (Schirmer, 2002a).

The investment companies have been responsible for a significant part of the increasing rates of plantation establishment during the past decade. However, their investments have predominantly established short rotation eucalyptus plantations, primarily as a consequence of the tax deductibility of the costs incurred in establishing and managing them. While this has attracted considerable investment from city-based investors, it has provided little incentive for the establishment of plantations by farmers on lower incomes who will not receive a significant tax benefit by doing so.

The role of the Commonwealth government in triggering investment in plantations can be seen by examining the recent changes in planting that resulted from changes in taxation rules in Australia. Before 2000, tax deductions operated under what was commonly referred to as the '13 month' rule.[8] As a result of a national review (the Ralph Review) of taxation in Australia, this rule was changed,[9] and a large increase in the area planted in the year 2000 resulted as plantation companies with irreversible commitments were forced to bring forward their planting to comply with the new policy (Australian Forest Growers, 2000). After the large increase in planting in 2000, investor uncertainty led to a downturn in establishment rates in 2001 and, when the Australian Tax Office (ATO) disallowed a large number of tax deductions for mass marketed agribusiness investment schemes, to a downturn in investment in all of these types of schemes, including plantation investment.

Plantations established by individuals and farm managers

The third form of plantation establishment on private land has been farm forestry – the establishment of a commercial tree crop on a portion of a farmer's property.[10] Since the 1970s, and occasionally earlier, various incentives have been put in place to try to encourage farm forestry throughout Australia. Principally, these have been grants and low-interest loans; funding of research and development; extension services; and development of joint venture and leasing programmes.

Grants and loans from state governments and agencies to encourage private planting have been offered in the past. However, these had a fairly low uptake as well as high administration costs, with some participants finding difficulties with repayments when they were unable to sell thinnings from their plantations (Hurley, 1986).

Funding of research and development of appropriate silvicultural regimes for growing plantations on agricultural land has occurred under government programmes, including the National Afforestation Programme (1988–1990) and the Farm Forestry Programme (1995 to present). Extension services have also been developed, such as the federal government funding and establishment of 17 regional plantation committees (RPC) through the national Farm Forestry Programme (Tuckey, 2000; NRRPC, 2001).

There is also a strong commitment by state and Commonwealth governments to foster a regulatory and policy environment conducive to farm forestry development under the *Plantations for Australia: the 2020 Vision* (MCFFA, 1997). Progress towards this goal was reviewed by Stanton (2000). Supporting initiatives have included policy and market research (e.g. Buffier, 2002; Race, 2002) and the provision of better market information (e.g. Bhati, 2002); but despite these efforts the expansion of farm forestry has been slow. The reasons for this are similar to the constraints[11] identified by Byron and Boutland's (1987) review of strategies to promote private timber production. It is notable that the most successful vehicles for encouraging uptake of farm forestry have been the joint venture and leasing schemes, precisely because they have addressed a number of these constraints.

Evolving demands and roles

Economic development: Commercial and social goals

The early dual motives for establishing plantations, under which governments and plantation proponents sought to generate regional economic development and employment through the establishment of both plantation forests and plantation-based industries, continue nearly a century later. For example, the *Plantations for Australia: the 2020 Vision* (MCFFA, 1997), developed jointly by the Australian Commonwealth and state governments and the forest industries, promotes the economic and employment generation outcomes of the expansion of plantation forests.

The importance of plantation forests in some regional and state economies remains strong. For these predominantly industrial commodity plantations, cost minimization and value recovery are central to the industry's future. Concurrently, there has also been a revival in public policy emphasizing the potential of plantations for broader regional economic development, and as a means of ameliorating environmental degradation in agricultural landscapes. However, enthusiasm for the social benefits of an expanded plantation resource have been tempered by concerns about the rural economic and social changes with which plantation forests are associated and, to varying degrees, held responsible by some community groups.

Other social issues

Attitudes to the form and scale of plantations in Australia have varied, from the great enthusiasm of the colonial acclimatization societies and the plantation-based industries to the strong concerns held by some rural and urban Australians about conversion of native forests to plantations; use of exotic monocultures (Routley and Routley, 1975); and social and economic impacts of industrial plantations (see, for example, Kelly and Lymon, 2000; Petheram et al, 2000, Tonts et al, 2001; Schirmer 2002a, 2002b).

Farmer groups and rural communities have expressed clear preferences for farm forestry-scale plantations. However, the majority of plantations established are of an industrial form and scale. This has been facilitated to some extent by the declining terms of trade for agricultural commodities and an ageing farmer population, both of which have encouraged some farmers to sell or lease their whole property rather than seek to integrate plantation forestry and agricultural production. Consequently, recent work (e.g. Williams et al, 2001; Buffier, 2002) has sought to identify policy and investment options which might also support farm-scale forms of afforestation.

Environmental benefits and costs

Over the past 30 years, community opposition to native forest harvesting has been matched by pressure from the environmental movement for the establishment of plantations to supply Australia's timber needs. Most Australian governments have been willing to facilitate this transition, up to a point; only Queensland has committed to phasing out wood production from native forests over a 25-year period. Similarly, opposition to the conversion of native forests to plantations has shifted the expansion of plantation forests entirely to previously cleared agricultural land, with the exception of Tasmania.

Many believe plantations should contribute more to biodiversity conservation and restoration, and have expressed a preference for multiple-species plantations and for the establishment of plantations of locally indigenous species and provenances (Schirmer, 2002a). Preliminary results from research suggest that plantations can be sited and managed to enhance biodiversity values, and that the additional costs of such management are not necessarily great (e.g. Catterall, 2000; Lindenmayer, 2000; Lindenmayer et al, 2002). However, despite these findings, the forest industries remain generally sceptical of the financial feasibility of mixed plantations on a large scale, while acknowledging their potential on a farm forestry scale and for particular high value products (e.g. Snell and Vise, 2000).

Other environmental concerns about the impacts of plantation expansion on cleared land have been expressed, including the effects of chemical use; the effects on soil and water quality and quantity; and the clearance of small pockets of native vegetation when plantations are established. These types of concerns appear to occur on a regular basis in different regions, indicating that the existence of significant regulation has not necessarily resulted in public confidence that plantation forestry is managed in an environmentally sustainable manner. This, perhaps, reflects the lack of broad consultation processes with communities. During recent years, there has been increasing consultation with a broader range of groups when undertaking activities such as reviewing codes of practice, which is necessary to help develop greater public confidence in forest practice systems.

Markets for environmental services

There has been considerable discussion of the potential for creating markets for the various environmental services provided by plantations, including markets for carbon, salinity and biodiversity credits (see Murtough et al, 2002; van Bueren, 2001). While these markets have yet to eventuate on any scale, there is strong public-policy interest in their potential to support plantation and plantation-based development (e.g. SFNSW, 2002; Williams et al, 2001), including in lower rainfall zones where salinity mitigation and biodiversity services from planted trees are usually most critical, and where returns from conventional markets alone are not commercially competitive. Some concerns prevail about emphasis on plantation carbon sink and salinity regulating functions where they may have negative impacts on water yields (Nambiar and Brown, 2001; Stirzaker et al, 2002). This increases in importance as competition for water rises in Australian agriculture.

Several states are preparing for the emergence of environmental services markets by enacting relevant legislation: for example, in May 2001, the Victorian government enacted carbon property rights legislation, which also supports investment in environmental plantings for purposes such as habitat expansion, mitigating salinity and land protection (Department of Natural Resources and Environment – Victoria, 2002).

Implications for public and private sectors

The evolving demands and roles described above are forcing a greater convergence between the behaviour of public- and private-sector actors in many aspects of plantation forestry. State plantation growers must now operate on a strong commercial basis, with less capacity to explicitly address social policy goals; conversely, community expectations of private industrial forest growers are expanding to deliver environmental services.

The roles of state governments with regard to plantation forestry have also become more strongly differentiated, with different agencies responsible for environmental oversight and regulation, for plantation growing, and for resource and industry development. There is now also greater diversity of private plantation owners, although industrial-scale growers and processors continue to dominate the plantation sector. These changes have been particularly rapid over the past decade, and it is arguable that the policy frameworks which might best balance public and private interests, nationally and in most states, are not yet fully developed.

Reconciling public-policy objectives and private-sector investment

Reconciling public-sector policy objectives and private-sector investment in plantation forestry

The principle Australian public-policy objectives relevant to plantation forestry over the past decade, and how they have been reconciled with private-sector investment, are addressed below.

Smaller government and more pervasive entrepreneurs

All Australian governments claim success in relation to these goals. They argue that the business performance of state forest agencies, and the business environment for private investment in plantation growing and processing, are better now than a decade ago. Most analysts and the plantation forestry industries agree, and this success is evident from the significant private-sector investments in Australia's plantation estate and processing capacity over the past decade. However, the prevailing arrangements advantage major growers and processors and disadvantage small-scale enterprise of a similar nature.

The public-policy challenge, which Australian governments have yet to successfully address, is how to balance the potential benefits of a vibrant and diversified small-scale forestry sector with the commercial realities of globally competitive industrial commodity wood production. So far, state governments have responded primarily through fostering a business environment conducive to large-scale industries, while supporting a range of national and state programmes which provide specific forms of assistance for small-scale growers – such as market intelligence and development, on-farm extension and partnership opportunities. However, these initiatives have not yet achieved significant expansion in small-scale commercial plantation forestry at a national scale. The development of markets and industries for products differentiated from those from industrial plantation forestry, and for environmental services markets, are widely seen as the best prospects for significant expansion in the small-scale forestry sector. Australian governments are playing active roles in both these policy arenas.

Smaller government and plantation forestry regulation

Diminution of regulatory impediments to plantation growing has been a major focus of public policy nationally and in all Australian states. While there has been good progress in a number of respects, the regulatory environment continues to vary markedly between states. Recent Australian experience with prospectus investment afforestation (e.g. Schirmer, 2002a) suggests that state and local government land-use planning and dispute resolution processes are not yet well adapted to addressing plantation afforestation.

The development and implementation of 'new generation' environmental regulation has been variable across states – reflecting lack of community consensus about acceptable forest practices. As a result, forest certification has been slow to emerge in Australia but now appears to be imminent, with some plantation growers undertaking the Forest Stewardship Council (FSC) process (e.g. Smartwood, 2002), the development of a national forestry standard (AFS Steering Committee, 2002), and the initiation of a complementary national FSC initiative (Cadman, 2002).

Shifting production from native to plantation forests

Australian governments have sought to manage the transition in commodity wood production from native to plantation forests principally through the *Regional Forest Agreement* process (CoA, 2000); the associated *Wood and Paper Industry Strategy* and its constituent *Farm Forestry Programme* (CoA, 2001); and the *Plantations for Australia: the 2020 Vision* (MCFFA, 1997) and related state initiatives.

In the few regions where there has been a significant uncommitted plantation resource, such as in south-west WA and parts of Tasmania, this transition has been fortuitously smooth in terms of resource supply, although often not in continuity of employment. Elsewhere, the geographical disjunction between native and plantation forest resources, and the differing investment and employment patterns of the two industries, have meant that regional economies and communities strongly dependent upon native forest production have not been able to derive much benefit from developments in the plantation forestry industries (e.g. CoA, 2000). Consequently, the aggregate national and state outcomes in the transition from native to plantation forest resources and industries, in fact, comprise both 'winning' and 'losing' communities within most states.

Renewed interest in fostering rural and regional development, and concurrent attempts to address rural environmental degradation

The declining terms of trade for traditional agricultural production, structural change in agriculture and Australian society, and environmental degradation have driven significant debate and policy initiatives during the past decade (see, for example, CoA, 1999; Gray and Lawrence, 2001). As discussed earlier, the development of new globally competitive industries, and the large-scale return of trees to the landscape, are widely seen as key elements of the response to these challenges.

Both traditional and innovative forms of plantation forestry are therefore important in this context and governments have sought to develop a variety of business and investment vehicles which combine public- and private-sector contributions. The most advanced initiatives developed on the basis of this partnership approach are those in WA, using both oil mallee eucalypts and drought-tolerant pines (Shea and Bartle, 1988; Williams et al, 2001) and the Murray–Darling Basin Commission's 'vegetation bank' (MDBC, 2002).

Public policy for these fundamental challenges facing rural Australia is based largely upon the capacity of governments to leverage private-sector capital and direct it in order to deliver public environmental and social benefits as well as competitive financial returns to the investor. This is an ambitious and desirable goal; but recent Australian experience with both prospective new industries and plantation forestry prospectus investments mediated through the tax system suggests that it is difficult to achieve.

The development pathway of relevant new industries suggests that high levels of both public investment and business and technological acumen are necessary to realize commercial crops which diversify farm incomes and address environmental degradation (e.g. Williams et al, 2001). The need for sustained involvement of governments in such endeavours should not be surprising, given their involvement in earlier phases of what are now established plantation industries, and especially as private investment capital is focused on a limited suite of financially competitive investments.

Lessons from Australia's experience

The Australian experience suggests the following guidance for increasing and sustaining private-sector participation while seeking to sustain the public interest:

Governments' roles may have changed, but still need to be strong, smart and coordinated

The roles of government have evolved significantly but are now best characterized as public business enterprise; business and investment facilitator and partner; and environmental and industry regulator. Australian experience suggests that if governments withdraw too far from these roles, or do not perform them well or in a coordinated way, market forces will not necessarily direct investment to locations or forms of plantation forestry which are in the public, as well as in commercial, interests.

All state governments other than Victoria's have preferred to retain state plantation forestry agencies within government, though as increasingly commercially focused government business entities. In some states these have entered into various forms of joint venture partnership with the private sector. The reasons that governments have chosen this intermediate option rather than that of full privatization reflect, to some extent, the concerns of many

Australians about the potentially adverse effects of privatization. Given the high level of commercial performance expected by state governments of the commercialized plantation growers, there may not be significant differences over the longer term in the behaviour of the commercialized state and privatized plantation growers.

The balance between public and private interest is a contested concept

The appropriate balance between public and private interest is, of course, contested. For example, the Australian experience with taxation incentives suggests that they can be a very powerful means of generating significant private-sector investment in some forms of plantation forestry; but it is also arguable that the rapid expansion of plantations they generated during the late 1990s was insufficiently tempered by appropriate land-use planning and conflict-resolution frameworks. Similarly, there is ongoing debate about the extent to which government agencies should enter into wood-supply agreements on favourable terms with regionally important processing industries; co-invest in plantation expansion for regional economic development and environmental services; and consider plantation crops to be treated as 'just another agricultural crop' under planning and regulatory regimes.

There are various forms of 'privatization'

Australia's federal structure of government, and its political and geographic diversity, has fostered a diversity of institutional arrangements for plantation forestry. It has also allowed the evolution of a spectrum of privatization, from the complete privatization of state-owned plantations to their corporatization, the development of a range of public–private partnerships in a variety of forms of plantations, and the encouragement of private- rather than public-sector investment. The Australian experience suggests that each of these has strengths and limitations: the closer to full privatization, the stronger the commercial imperatives, the greater the access to private capital, and the more sophisticated the public policy instruments required to deliver public good outcomes; conversely, the further from commercial orientation, the greater the need for ongoing public-sector investment, but the more easily that investment is directed to meet public policy goals.

Australia's experience of privatization outside the forestry sector suggests that neither unfettered privatization, nor retention of highly regulated quasi-government entities, is likely to be optimal where a balance of public good and economic efficiency outcomes are sought. It is probably too early to draw any conclusions in these terms from the relatively recent experience of plantation privatization in Australia; but the various models of privatization reflect different state governments' judgements of how that balance is best struck, and offer fertile ground for future studies as the various models evolve and their consequences become clearer.

Various forms of public–private partnerships can work well for all parties

Australian plantation forestry is characterized by a variety of forms of public–private partnerships, from joint policy initiatives and the development of joint investment products, to on-ground joint ventures between state forest agencies and landowners, and collaborative research and development programmes. As the interests and behaviour of state forestry agencies and corporate plantation growers have converged, so has their capacity to cooperate as well as to compete. In this realm, as in others, systems based on partnership appear to work better than those which are adversarial. Both governments and the private sector also need to find vehicles, such as certification, for partnership with the community and with environmental non-governmental organizations (ENGOs) in order to ensure the community support necessary to sustain plantation forestry.

Delivery of non-commercial and non-market benefits from commercial plantation forestry is a particular challenge. Debate continues about the appropriate form and level of cost-sharing between public and private interests to achieve delivery of non-commercial benefits by private plantation growers. The Australian experience suggests that a diversity of forms of public–private partnership is necessary to respond to the particular economic, social and environmental dimensions of different states and regions.

Engaging small-scale growers fully in plantation forestry is particularly challenging

Small-scale prospective plantation forest growers in Australia face many obstacles in addition to those faced by industrial-scale forest plantation growers (e.g. diseconomies of scale, limited access to capital and more diverse objectives). Australian plantation and farm forestry policy has sought to address these for nearly 20 years, with only modest results. Policy has succeeded in fostering much more of a tree-growing culture among farmers; but the markets required to drive reforestation on a large scale have yet to emerge for more than a small number of products. Innovative attempts to address these challenges, particularly in forms that integrate tree-growing with traditional agricultural production, demonstrate both exciting potential and significant constraints which limit the development of new industries. A strong partnership role for both the public and private entities is necessary if the participation of small-scale growers in various forms of plantation forestry is to be fostered to a point where it can be self-sustaining.

The best mix of regulation and incentive for private forestry will vary with contexts

The diversity of regulation and incentive structures relevant to private plantation forestry across and within the Australian states offers clear evidence that some forms of incentive and regulatory systems work better than others. While there are principles which apply generally to both incentives and to regulation, there is also the need to adapt policy frameworks to the particular community, business and political environments of plantation regions.

Recent Australian experience with prospectus-based investment in plantation forestry suggests that one of the strongest incentives for private-sector plantation forestry investment is to ensure that it suffers no more obstacles than do investments in alternative land uses. Similarly, in terms of removing distortions from investment and land-use decisions, forest practices systems should apply similarly to both public and private forestry, and appropriate environmental standards should apply to agriculture land-use alternatives, as well as to plantation forestry.

Conclusions

Australia's governments are intent on continuing down the established path of the past decade, towards maintaining 'small government' and facilitating both domestic and international investment in the Australian plantation forestry sector. However, most have preferred to retain state forestry agencies within government as quasi-commercial trading enterprises, than to privatize them fully. Concurrently, Australian governments have encouraged and enabled the private sector to expand Australia's plantation forests through favourable taxation arrangements for plantation investors, and through facilitation of a diversity of forms of public–private partnership in plantation forestry. These have been more successful with large-scale than with small-scale plantation forestry, although the latter has been and remains a political priority.

The Australian experience with plantation development, suggests that the more commercial orientation of corporatized government plantation management has generally been in the public interest, although there have been adverse impacts, for example, on direct employment. Whether or not full privatization becomes more common, governments will need to continue to develop more sophisticated policy regimes and instruments to deliver many public policy goals from a more commercial, more widespread, and more important plantation forestry sector.

6

Early Experience of Total Divestment: Chile

Eduardo Morales

Geography and demography

Mainland Chile covers the south-western coast of South America (see Figure 6.1), but 62.3 per cent of its total area is in the Antarctic. Most activity takes place on the mainland, where nearly all of 15.4 million Chileans live.

The country has 13 regions, 51 provinces and 355 municipal districts. Forestry activity takes place from Region V, de Valparaíso, south to Region XII, de Magallanes y de la Antártica Chilena, where it is typically Mediterranean, with cold, wet winters and warm, dry summers. Further south the rainfall increases and there are lush native forests and forest plantations.

The Atacama Desert covers nearly one third of the country, from Perú to La Serena in the centre. Another third is either broken land split into thousands of small islands or continental ice that stretches from just South of Chaiten to Southern Tierra del Fuego. The remaining 1300km is Central Chile, where most forest and agricultural land is found.

Region VIII (Concepción) is the centre of the country's forest activity. Its abundant plantations sustain pulp and paper manufacturing, saw mills, and the production of wood panels. Regions IX (de la Araucanía) and X (de Los Lagos) are the centre of native forest activity, and as well as sawn native timber they produce wood panels, such as veneers and plywood.

Chile has one of the lowest population growth rates in South America. Between 1970 and 2001, it grew by an average of 2.6 per cent per year. Nearly 40 per cent of the population lives in or near the capital, Santiago, in the metropolitan region.

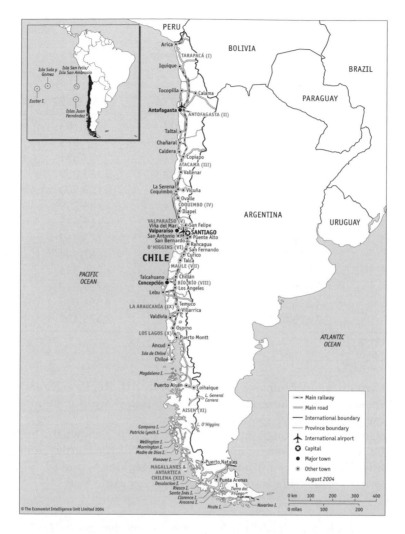

Figure 6.1 *Chile: A long strip of land*

The forestry sector: Some basic facts

Some 20.6 per cent (15.6 million hectares) of Chile's territory (excluding Antarctica) is covered by forests (see Table 6.1). Of this area, 6 million hectares are native production forests (all privately owned), 7.5 million hectares are protected forests (5.3 million hectares owned by the state and 2.2 million hectares by private owners) and 2.1 million hectares are plantations (all private).

Table 6.1 *Total forest area of Chile by regions (hectares)*

Regional distribution	Native forest productive and protected	Plantations	Total forests	
	13,443,316	2,118,836	15,562,152	100
I	7682	28,502	36,184	0.23
II		674	674	0.00
III		1891	1891	0.01
IV	1377	61,682	63,059	0.41
V	94,008	64,892	158,900	1.02
RM	93,345	13,214	106,559	0.68
VI	117,798	97,825	215,623	1.39
VII	369,708	403,652	773,360	4.97
VIII	785,766	882,178	1,667,944	10.72
IX	907,521	351,475	1,258,996	8.09
X	3,610,314	187,514	3,797,828	24.40
XI	4,830,743	25,289	4,856,032	31.20
XII	2,625,054	47	2,625,101	16.87

Source: INFOR, 2001

Most native forest area remains in Regions X to XII. Regions XI and XII are almost inaccessible because of the broken land and steep terrain, but have some slow-growing native species plantations.

Plantations exist throughout the country. Table 6.2 shows the breakdown of area by species. There are industrial plantations of (primarily) radiata pine, eucalyptus, Douglas fir and poplar between Regions V and X. The north mainly grows atriplex, blackwood and prosopis for cattle feed.

Table 6.2 *Area of Chile's plantations by species (2001)*

Species	Area (ha)
Radiata pine	1,570,058
Eucalyptus	381,390
Atriplex	57,209
Prosopis chilensis	23,307
Douglas fir	14,832
Poplar	4238
Schinuss molle	3505
Other species	64,298
Total	2,118,836

Source: INFOR, 2001

Table 6.3 *Commercial forests in Chile (2001)*

Forest type	Area (hectares)	Percentage
Native forest	13,443,316	**85.91**
Adult	5,978,200	38.20
Second-growth forest	3,582,408	22.89
Adult/second growth	865,446	5.53
Krummholz	3,017,262	19.28
Plantations	**2,118,836**	**13.54**
Mixed forest	**85,742**	**0.55**
Total	**15,647,894**	**100.00**
Native forest	13,443,316	**85.91**
Commercial	5,978,200	38.20
Non-commercial	7,465,116	47.71
Plantations	**2,118,836**	**13.54**
Mixed forest	**85,742**	**0.55**
Total	**15,647,894**	**100.00**

Table 6.3 shows the total commercial forest area of Chile. Native forest is the largest category, but contributes only 6 per cent of industrial logs because radiata pine products are cheaper.

Legal and policy context for forestry

Two laws govern the most important aspects of forestry, the updated Forestry Law of 1931 and the Forestry Development Law (DL 701) of 1974 and its subsequent modifications, regulations and resolutions.

Forestry Law of 1931

Forestry Law No 4363, passed on 30 June 1931, deals with the environmental aspects of forest operations. Articles 1 and 2 define forest lands. Article 3 could be the first attempt to promote plantation forestry in Chile as it says: 'The existing artificial forests, or those to be planted on declared forest lands, will be tax exempt for a period of 30 years.' This article is still in force with respect to plantations established before 1974.

Article 5 is the most important from an environmental standpoint. It is still enforced and dictates that no trees can be felled within 400m of a stream or riverbank or within 200m of their source. There are a further 16 articles that have been modified or expanded by subsequent decree laws or laws.

Forestry Development Law DL 701

Decree Law No 701 is the most important regulatory act and governs natural and planted forest use in Chile. The law and its subsequent amendments require coherent forest management and empower the National Forestry Corporation (Corporación Nacional Forestal, or CONAF) to create management norms for specific forest types to ensure sound professional management.

The main aspects of DL 701 are as follows:

- The classification of lands as 'suitable for forestry preferentially' is the basis upon which the whole incentive structure is built. The owner submits to CONAF a technical study of the land and a reclassification proposal.
- The Forest Management Plan is also embedded in DL 701. If land is reclassified, the owner must present to CONAF a management plan prepared by a forestry engineer or a specialist agronomist. The plan must detail all afforestation within a period of the first five years, and reforestation three years after felling.

Other forestry laws

Other acts regulating the forest sector include acts to protect and control logging, to protect endangered species, and to prevent and control forest fires. A summary of this legislation is available at www.conaf.cl.

A commission directed by the Undersecretary for Agriculture is preparing a native hardwood policy. This project has been in the Chamber of Representatives for 12 years and is still under discussion, having been withdrawn and resubmitted by CONAF several times. Clear policies regarding the management, use, logging and substitution of native hardwood forests will be determined on the basis of this law and ground rules for additional forestry logging and development will be established.

Environmental laws

Law No 19,300 establishes a system of environmental impact assessment that applies to major industrial-scale projects in all sectors – not just forestry. The basic law was passed in 1994, but implementing regulations pertaining to environmental impact assessments were not promulgated until 1997. CONAMA, the National Environment Commission (Comisión Nacional de Medio Ambiente), is responsible for evaluating individual impact assessments. A major investment proposal for a natural forest area was recently halted on the basis of these regulations.

Protected areas, water and wildlife

Chile is strongly committed to sustainable development. It has signed the Convention on Biological Diversity (CBD), so conservation measures that underpin projects and forest management activities must comply fully with Chile's obligations under the CBD.

Ecologically interesting protected wildlife areas make up the State Protected Wildlife Areas System (SNASPE), owned by the government and administered by CONAF. SNASPE's 14.1 million hectares are divided into three categories: national parks (62 per cent of the total), national reserves (37.9 per cent) and national monuments (0.1 per cent) (see Table 6.4).

Table 6.4 *Chile's national parks and forest reserves by regions (hectares)*

Region	National parks	National reserves	Natural monuments	Total State Protected Wildlife Areas System (SNASPE)
	8,718,260	5,387,433	17,880	14,123,573
I	312,627	309,781	11,298	633,706
II	268,671	76,570	31	345,272
II – III	43,754	–	–	43,754
III	104,790	–	–	104,790
III – IV	–	859	–	859
IV	9959	4229	128	14,316
V	24,701	19,789	5	44,495
RM	–	10,185	3009	13,194
VI	3709	42,752	–	46,461
VII	–	18,669	–	18,669
VIII	11,600	72,759	–	84,359
IX	147,538	149,022	172	296,732
X	491,324	112,716	2517	606,557
XI	2.064,334	2,223,913	409	4,288,656
XI – XII	3,525,901	–	–	3,525,901
XII	1,709,352	2,346,189	311	4,055,852

Source: INFOR, 2001

Key incentives for private plantations

An owner who establishes plantations under DL 701 has the following benefits:

- access to planting and management subsidies if the land is classified as preferentially suitable for forestry and the owner is a small property owner;
- exemption from the land tax applied to agricultural lands until two years after the first rotation;
- exemptions from inheritance tax;
- profits from the use of forests, which attract additional income tax benefits.

When CONAF extends planting subsidies, the owner has the following obligations:

- To submit a technical study of the land and a classification proposal.

- Within a year of the land being reclassified the owner must present a management plan to CONAF. CONAF has 120 days during which to object to the management plan. If it does not, approval is assumed.
- Logging is only allowed after CONAF has approved the management plan. Breaking this rule will make the owner or person who carried out the unauthorized felling liable to a fine of twice the commercial value of the products as determined by CONAF, and the products are confiscated if unsold. If sold, the transgressor will be fined three times their commercial value. CONAF can order immediate stoppage of work and may use the police if necessary.
- The owners must reforest or recover an area at least equal to that felled or logged, although not necessarily the same land if an alternative is approved by CONAF. The resulting plantations will be considered reforestation for all legal purposes.
- If reforestation does not occur within three years after felling, owners will be subject to fines of 10 to 30 *unidad tributaria mensual* (UTM) per hectare, which in July 2002 amounted to US$415 to US$1248 per hectare. The higher end of the scale is about double the cost of reforestation.

The forestry classification may be annulled in exceptional cases. The owner must then pay back any subsidies or tax exemptions, plus adjustments and interest determined by the Internal Revenue service.

Key actors in the forestry sector

The public and private sectors play quite different roles. The public sector controls and enforces laws and provides and administers promotion and subsidy tools. The private sector is responsible for production and provides support services, such as finance, ports, transportation, marketing, legal consulting, accounting, maintenance, and so on.

Public administration

Corporación Nacional Forestal (CONAF). The National Forestry Corporation, under the Ministry of Agriculture, manages national forestry policy. Its main objective is to contribute to the preservation, management and use of the country's forestry resources. CONAF has two main functions:

1 CONAF ensures that the forest industry applies and complies with the law, norms and regulations regarding forest management, logging and other operational issues.
2 CONAF manages SNASPE (Sistema Nacional de Areas Silvestres Protegidas), the National System of Protected Wildlife Areas.

Servicio Agrícola y Ganadero (SAG). The Agriculture and Livestock Service oversees compliance with legal requirements for seeds, fertilizers, pesticides, flora and fauna and for sanitary detection and control, particularly of both imported and exported natural products.

Instituto de Desarrollo Agropecuario (INDAP). The Agricultural Development Institute provides financial assistance to small- and medium-sized farmers and to organizations developing activities in low-income rural areas.

Ministry of the Economy. This ministry promotes and looks after the development of the country's various productive and service activities and houses CORFO (Corporación de Fomento de la Producción).

Comité de Inversiones Extranjeras: The Committee for Foreign Investment approves and controls the entrance of foreign capital and promotes foreign investment, provides legal and financial assistance to foreign investors, and aims to treat all investment applications equally.

Corporación de Fomento de la Producción (CORFO) The Industrial Development Corporation's main purpose is to promote the development of productive sectors. It administers sectoral development initiatives, including grants for research and development by universities, and private and public research organizations.

Among the private- and public-funded agencies supported by CORFO, the following are forestry related:

- *Instituto Forestal (INFOR)*: the Forestry Institute improves the development of the forestry sector through research and technical assistance.
- *Instituto de Investigaciones Tecnologicas (INTEC)*: the Technical Research Institute provides technical support to industry and CORFO. This involves creating, introducing, adapting or changing technological processes through the use of technical information, technology transfer and scientific testing.
- *Servicio de Cooperación Técnica (SERCOTEC)*: the Technical Cooperation Service helps small- and medium-sized industries to increase their productivity.

Private institutions
Companies, independent professionals and concerned people with ties to the forestry sector are grouped within the following organizations:

- *Corporación Chilena de la Madera (CORMA)*: the Chilean Wood and Forest Products Association gathers together producers, professionals and institutions connected to the forestry sector. Its main purpose is to promote the development of the Chilean forestry sector and the optimal use of forest resources.
- *Asociación de Industriales de la Madera (ASIMAD)*: the Wood Re-Manufacturers Association brings together producers of value-added products, such as doors, windows, furniture and furniture components, among other products.
- *Asociación Técnica de la Celulosa y el Papel de Chile (ATCP Chile)*: the Pulp and Paper Association of Chile is a group of professionals working in pulp and paper companies in Chile. Its main purpose is to improve the professional and technical skills of its members by promoting technological development and scientific research.

- *Colegio de Ingenieros Forestales*: the Forestry Engineers Association is the society of professional forestry engineers. Its purposes include promoting cooperation among members; fostering technology innovation; disseminating scientific information; and contributing to community development.
- *Fundación Chile:* this non-profit, private institution was created in 1976 by an agreement between the Chilean government and the Corporation of the US ITT corporation to transfer technologies that contribute to the rational use of natural resources and to the country's production, mainly by creating new enterprises.

There are few non-governmental organizations (NGOs) in Chile. Two leading environmental NGOs are *Defensores del Bosque Nativo* (Native Forest Defenders) and CODEFF (Flora and Fauna Defence Committee). There is a third, academic, NGO, CIPMA (*Centro de Investigación y Planificación del Medio Ambiente*). Besides these three, there is a group of professionals calling themselves *Asociación Chilena de Ingenieros Forestales por el Bosque Nativo* (Association of Chilean Forestry Engineers for Native Forests). The Native Forest Defenders, along with Forest Ethics of the US, have recently started a campaign to prevent the export of radiata pine forest products not bearing the Forest Stewardship Council (FSC) logo.

Forestry tenure systems

Since 1974, the country's forest-based industry – the commercial forests and the processing industries – have been private. That year, the socialist government of Salvador Allende was replaced after a coup by a military Junta headed by General Pinochet. The new government declared that it would prevent the state from carrying out business activities and would promote free enterprise. As a result:

- Two of the three pulp mills owned by CORFO were privatized.
- Nearly 120,000ha of a series of large forestry properties that had been expropriated by the socialist government for agrarian reform were restored.

CONAF continued to establish plantations until 1978. Between 1983 and 1985, CONAF initiated a planting programme aimed at combating unemployment.

All native forests outside SNASPE are privately owned. Those on steep terrain are treated as protected and cannot be harvested, even though privately owned.

Of the 15.65 million hectares of forests, 5.8 million hectares are within SNASPE and 9.85 million hectares are private sector.

Of the total planted area, a little more than the half is owned by eight major forest companies, the rest, mainly in Region V, by several thousand small- and medium-sized owners (see Table 6.5). Plantations of over 1000ha account for

Table 6.5 *Plantation ownership in Chile by size of property*

Property range size (hectares)	Number of owners	Area (hectares)*	Percentage
Large companies	8	1,097,000	52.9
Individual owners:			
Over 2000	195	281,328	13.6
1000 to <2000	140	196,019	9.6
500 to <1000	230	157,198	7.6
200 to <500	451	139,527	6.7
100 to <200	566	79,270	3.8
50 to <100	831	58,762	2.8
10 to <50	2130	51,947	2.5
1 to <10	2880	10,086	0.5
Total	7431	2,071,137	100.0

Note:
*Species: Radiata pine, eucalyptus, poplar, blackwood, Douglas fir, native.

Source: adapted from INFOR, 1997

75 per cent of the plantation estate, and they are owned by only 4.6 per cent of plantation owners – a very high concentration.

Economic significance

Contribution to GDP

In 1980, the total value of forestry production was US$774 million. By 1994, the sector was the fifth largest contributor to gross domestic product (GDP) at US$2 billion. Its contribution to GDP during the same year represented 3.18 per cent of the domestic total (assets and services), or 7 per cent of the gross national products (GNP) assets, up from 2.4 per cent GDP in 1975.

In 1994, the main sub-sectors of forest GDP included silviculture and wood extraction (18.6 per cent), wood products (25.3 per cent), and furniture and pulp and paper (56.1 per cent). By 2000 these shares had changed to silviculture and wood extraction (12.4 per cent), wood products (35.4 per cent), and furniture and pulp and paper (52.2 per cent). The furniture industry has grown most rapidly during recent years, surpassing the rate of growth of the pulp and paper industry. Forestry's contribution to total GDP has declined since 1995, mainly as pulp prices deteriorated, although sectoral GDP has increased by 7.4 per cent.

Forest-based industry

The primary forest-based industries produce lumber, wood-based panels and pulp products, while the secondary industries remanufacture lumber for furniture production and produce remanufactures, such as mouldings, cut stock, blanks, and furniture and its components. The secondary paper industry processes pulp for newsprint and other papers, cardboard and paper boxes. The flow of industrial radiata pine logs is shown in Figure 6.2.

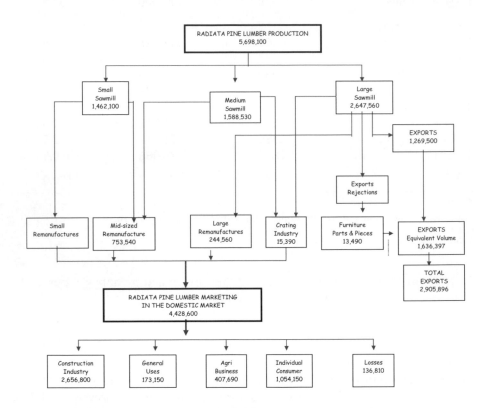

Figure 6.2 *Flow of industrial logs in the Chilean forest industry (cubic metres in 2000)*

Currently, 93 per cent of industrial wood production comes from plantations. Within eight years, that is expected to be almost 100 per cent as more pine and eucalyptus plantations come on-stream and logging native forests becomes increasingly difficult. Short-fibre wood chip exports to Japan from native species will be completely replaced by large eucalyptus plantations in Region X and the north.

Controlled afforestation with radiata pine has been the basis of considerable industrial growth. For 30 years, annual growth of radiata growing stock has exceeded 6 per cent and this will probably continue until 2025.

By 1994, annual planting had reached 109,900 hectares, but thereafter decreased as land became expensive and larger companies found northern

Argentina, southern Brazil and Uruguay more profitable. The rate of planting is expected to recover. An additional 2.5 million hectares of privately owned land is suitable for plantations, so there is no pressure on protected areas.

By 2007, vast areas of planted forest are due to mature, providing an increased supply of saw logs, peeler logs and pulpwood. This new supply will likely be cheaper and more uniform, which will promote additional industrial capacity.

Forestry attracts both domestic and foreign investment. Foreign investors are generally from countries with a forest tradition, and contribute experience in both sustainable forest management and modern technology.

Exports

The Chilean policy of low and uniform tariffs has increased exports. Since 1974, forest exports have grown hugely, in both value and volume (see Table 6.6). Forest products' export share has increased steadily, peaking at almost 15 per cent in 1995, but remaining the same since because of reduced prices in pulp and remanufactures.

Table 6.6 *Exports of Chile's forest products as a percentage of total national exports*

Year	Total exports	Forestry exports	Share (percentage)
1970	1111.7	41.7	3.8
1975	1552.1	125.5	8.1
1980	4670.7	468.1	10.0
1985	3823.0	334.6	8.8
1990	8580.3	855.3	10.0
1995	16,039.0	2369.3	14.8
2000	18,158.0	2365.2	13.0

Source: Estadísticas Forestales; INFOR, 2001

Table 6.7 shows the progress Chile has made in diversifying its exports market, in both trade partners and products, particularly in value-added products from the wood remanufacturing industry.

Forests and the poor

The national forest policy does not directly or primarily aim to alleviate poverty, promote small enterprises or subsidize employment; all of these outcomes are expected to happen as a result of the growth and industrialization of forest capital and the marketing of forest products worldwide.

During the last three decades, forestry and related industries have not directly significantly alleviated unemployment or poverty in the plantation

Table 6.7 *Diversification of Chile's forestry exports*

Item	1966	1980	1999
Value of forestry exports (million US$)	20	580	2365
Number of countries of exports' destination	12	32	98
Areas of destination (percentage)			
South America	75	45	18
North America	20	5	28
Europe	5	16	18
Asia	–	34	36
Number of products	40	185	462
Type of products (percentage)			
Logs	–	11	5
Lumber	70	51	36
Wood-based panels	3	2	5
Pulp	27	36	37
Clear blocks	–	–	3
Finger-jointed material	–	–	1
Edge-glued panels and planed lumber	–	–	13

Source: INFOR, 2001

areas. However, indirect employment has exploded, not only for forest labourers, but also in related activities such as transport, logging and equipment maintenance. These new jobs are better paid, and the affected regions are no longer the poorest.

Neither CONAF nor the Ministry of Agriculture include explicit concerns about poverty in their policies and activities. This may be because Chile is mostly an urban economy: 86 per cent of people live in cities and most of the rural population live in agricultural areas, with only a small proportion in forest areas. In Chile, urban poverty is a far bigger issue than rural poverty.

A brief history of state-owned plantations

The origins of forest plantations in Chile

Around 1865, the first foreign exotic ornamental eucalyptus trees, mainly *Eucalyptus globules,* were introduced to Chile, mostly planted in rows along paddocks in the agricultural regions of central Chile. In 1885, Arturo Junge of Concepción started planting North American conifers. He found that radiata pine grew much faster than Douglas fir, so he planted it extensively as an ornamental. Between 1907 and 1912 Konrad Peters planted 400ha of radiata for pit props for coal mines – the first really industrial plantation in Chile. In 1915, several landowners followed with radiata and other species, including poplar.

By 1930, 0.5 per cent of the industrial raw material used by the forest industry was radiata pine. Most wood still came from native forests; during 1920–1930, more than 2.5 million hectares were cleared. The Forestry Law of 1931 protected native forests and encouraged afforestation by exempting forest landowners from land and inheritance taxes while offering incentives for afforestation. Around 1925, there were between 5000 and 6000 ha of radiata; by 1929 the Lota Mining Company had 7300ha of both radiata and eucalyptus.

CORFO (the state-owned development agency) was created by the socialist government of Aguirre Cerda in 1931. In 1940, CORFO initiated a vast private plantation programme supported by subsidized loans. The annual rate of plantation establishment – all private – grew from 1602ha in 1929 and peaked at 49,257ha in 1949, 90 per cent radiata. By 1964 the annual planting rate had declined to only 13,176ha as private investors were worried about political instability and land reform processes.

Notwithstanding this, several saw mills, a pulp mill and a fibre board mill were established and a market for logs was developed. Thousands of small farmers started planting, hoping to later sell their timber to these industries.

State-owned plantations

It was not until 1970, nearly 85 years after the first plantations were established, that the state became involved in plantation forestry. Between 1970 and 1985, the state's estate grew from 6940 to 365,729ha.

Plantation privatization
In 1973, a coup deposed Salvador Allende's socialist government and the new rightist Junta radically changed the way in which the economy was managed. The Junta had pledged to keep government out of business, and in 1974 they passed Decree Law 701, a cornerstone of Chile's plantation policy, with an emphasis on private-sector planting but with huge state subsidies. The basis of this law is as follows:

- *Prerequisites:* land ownership; an approved forest management plan; and land declared as of 'preferred forestry aptitude'.
- *Benefits:* a 75 per cent subsidy for afforestation costs, based on a cost table published by CONAF six months ahead of the planting season; a variable management subsidy, depending upon the type of management applied, plus a subsidy to cover administration costs; and exemption from land tax. Fifty per cent reduction on integrated income taxes for individuals and corporations.
- *Commitments:* the status of the land cannot be changed from 'preferred forestry status' unless it is replaced by an identical area elsewhere; reforestation is mandatory a year after harvest, or else heavy fines and possibly land confiscation result.

Decree Law 600 of March 1977, which promoted foreign investment, was instrumental in promoting forestry-sector development, attracting several multinational companies.

In 1979 CONAF still controlled 32 per cent of the total area planted in Chile. A serious economic recession during 1983 to 1985 held up privatization, while the government, through the municipalities, launched a new plantations programme to provide jobs.

Following the recession, CONAF stopped planting and sold or transferred its remaining holdings to private entities – some to the landowners, others to large companies.

From 1986, afforestation and reforestation rates grew steadily and peaked in 1992 at 130,429ha per year. Thereafter, plantations rates fell (see Table 6.8).

The rate of planting fell steadily after 1992, mainly because:

- Land became expensive compared to Uruguay and Argentina.
- As available properties became smaller, transaction costs per property went up.
- Land was more productive in Uruguay and Argentina, so both large companies and small investors moved.

Conversion of native forests to plantations

Most new plantations have been developed on degraded forest land or barren land. Only 0.4 per cent of the total planted area was native forest, most of which is second growth. DL 701 subsidies do not apply to previously forested areas in order to deter conversion. Many millions of hectares of degraded agricultural land and extremely degraded forest land are still available to industry for afforestation.

Evolving demands and changing roles

Concerns about plantation forestry

The main concerns about monoculture plantation forestry are disease and loss of biodiversity. Environmental groups would like to see an increased use of native species in plantations and more diversity of exotic species, but the financial and economic implications are not yet known.

Harvesting forests and building roads can be hazardous for water supplies; but harvesting must be approved by the government and management plans include notices of intended cutting.

Harvesting operations in plantations are carried out using a combination of oxen, skidders, cable yarders and bunchers, causing very little erosion. Since the companies own the land, there is sufficient incentive to maintain the quality and fertility of the land.

Table 6.8 *Chile's historical planted area by the state and private sectors, 1970–2000*

Year	State	Private	Total
1965		18,660	18,660
1966		40,618	40,618
1967		37,070	37,070
1968		32,328	32,328
1969		35,908	35,908
1970	6940	16,506	23,446
1971	16,659	11,387	28,046
1972	24,772	6270	31,042
1973	27,403	2910	30,313
1974	35,196	21,027	56,223
1975	44,096	38,463	82,559
1976	54,170	53,635	107,805
1977	44,673	48,499	93,172
1978	24,885	52,486	77,371
1979	477	51,749	52,226
1980	85	72,079	72,164
1981	29	92,753	92,782
1982	41	68,545	68,586
1983	21,811	54,458	76,280
1984	40,302	53,300	93,602
1985	24,190	72,084	96,274
1986		66,197	66,197
1987		65,441	65,441
1988		72,508	72,508
1989		86,703	86,703
1990		94,130	94,130
1991		117,442	117,442
1992		130,429	130,429
1993		124,704	124,704
1994		109,885	109,885
1995		99,858	99,858
1996		78,592	78,592
1997		79,484	79,404
1998		86,579	86,579
1999		108,269	108,269
2000		102,350	102,350
Total	**365,729**	**2,303,306**	**2,669,035**

Source: INFOR, 2001

What people want from plantations

Timber is only one product from plantations – they are also valued for their non-timber forest products (NTFPs) and environmental services. Radiata pine, in particular, has restored eroded soils and controlled run-off on steep terrain. Arguably, they could enhance some degraded scenery, and the Native Forest Defenders claim that plantations are not forests and that radiata pine forests are 'green desert'.

Plantations have also helped to protect natural forests by providing an alternative source of raw material. The illegal cutting of native forest by local people and small companies continues. They may be cutting a large proportion of more than 12 million cubic metres per year, mostly for firewood.

Changing roles

Most of the few NGOs in Chile are concerned with native forests, while plantations are generally seen as 'production forests' and, as such, are not expected to render the environmental services demanded by the public.

While production is now in the private sector, the state regulates and controls these activities and provides 'forestry services' through SNASPE. The public seems pleased with what CONAF is doing in these areas. If it is considered that 14.1 million hectares, out of 75.7 million hectares of the continental territory, are under the management of the state through CONAF as part of the SNASPE, this means that 18.6 per cent of the territory is devoted to the supply of forestry-related services from protected areas, among the highest in the world.

Changing roles of the state

Until 1960, all land was privately owned, sometimes as part of a farm far too big for the owners to work, so they relied on hiring local labourers, who were paid little and often had no land of their own. This began to change, and between 1960 and 1973 Chile experienced several waves of land reform. The government's role changed from one of limited regulation to a very interventionist one, and much land was expropriated and handed over to landless people or turned into Soviet-style giant farm businesses, complete with their own support systems of schools, clinics and shops. During 1964–1970, under the Christian Democrat government, more land was expropriated than in the previous 20 years, minimum wages in agriculture became the same as for industry, family allowance to farmers was doubled, and 44,000 small landowners received supervised agricultural credit, compared to only 12,000 the year before.

Direct redistribution of land peaked under Salvador Allende. From April 1972, the Popular Unity coalition movement expropriated 2678 farms covering 4.25 million hectares. They also created the state-funded and administered COREF, the Corporación de Reforestación (Reforestation Corporation), which started a wide afforestation programme on 'reformed lands' (expropriated land that now belonged to agrarian reform co-operatives). In 1972, COREF was transformed into CONAF (Corporatión Nacional Forestal), with a wider mandate to afforest and reforest on a national scale.

The government also helped two somewhat backward areas – Arauco and Constitución – by building a large CORFO-owned pulp mill in each. The raw material would come from existing plantations, including the recently established COREF plantations, and the mills would decrease the unemployment and underemployment caused partly by over-farming and overgrazing during the past century and, to an extent, by the consequences of the agrarian reform.

At the same time, the government expanded the plantation estate through Convenios de Reforestación (Reforestation Agreements), through which the owner of a reformed property (a property expropriated under Agrarian Reform and awarded to a peasant or new owner) contracted with CONAF to provide the land if CONAF provided the resources for planting. The owner managed the crop and then split the profits with CONAF at harvest. Thus, there were two state-owned plantation programmes: one on land purchased or leased from private owners with the crop wholly owned by the state, and Convenios, in which CONAF held rights to a share of the profits.

After the coup in 1973, Chile experienced a 'counter-agrarian reform', when the military Junta reallocated land to those with the technical skills, the will to work the land and access to capital.

Reconciling public-policy objectives and private-sector investment

Privatization and the stated government goals

All plantations are now privately owned and operated. The government is overseer and regulator. A constitutional amendment would be needed to renationalize any land, and that is very unlikely. If the state wants to establish new parks or reserves, constitutional procedures are required to do so.

Chile encourages free enterprise and free trade, and has simplified bureaucratic procedures and has a good labour force. Current trends include innovation in export market products; the application of new technology to industrial processes; free competition in local industries; development of joint private-sector ventures; foreign participation in technologically oriented industries; and expanding service-oriented industries.

Government policy aims to maintain the highest possible economic growth rate, low inflation rates and realistic foreign exchange rates.

The emphasis is on promoting private-sector involvement in social expenditure to improve education, housing and health services. Capital expenditure on social needs is leveraged through concession agreements. Ports and highways have been privatized, as have water catchments and distribution, and electricity generation and distribution. Social and pension funds are administered by private companies and education has been privatized through the municipalities. There is a three-tier healthcare system: private, part-privatized through municipalities and a few public hospitals.

Decree Law No 600 of March 1977 (known as the Foreign Investment Statute), promotes foreign investment to provide capital and technology for economic development. In 1985, Chile even withdrew from the Andean Common Market (ANCOM) in order to be free from ANCOM's foreign investment restrictions. Practically all forms of business activity are open to foreign investors and there are no ownership restrictions. Some regional investment incentives are available. Activities fundamental to the country's development, such as small mining, transport and forestry, also have special incentives. Capital and technology are sought to develop new sources of copper, nitrates, iron, coal and petroleum, as well as for manufacturing, forestry, agriculture and fisheries.

A continuing role for the government

CONAF issued a document laying out the government's role in forest policy (CONAF, 2000). The main institutional objectives are to:

* Contribute to the growth and sustainable use of the forest resources by developing tools to promote new plantations, managing the existing forests and promoting forest research.
* Conserve the representative natural ecosystems of the biological diversity of Chile.
* Help improve the quality of life of rural people through forestry activities by assisting in the commercial management of properties with degraded soils, promoting the introduction of more valuable species and promoting equal opportunities for rural women.
* Protect the forest ecosystems from dangerous pests through the development of environmental, economic and socially acceptable tools, the integrated management of watersheds to strength the relationship between soil, water and vegetation, and the promotion of integrated resource management.

Expectations and government policy

Chile was a traditional democracy, which came to an abrupt end in 1973. For the next 27 years, the country was ruled by the dictator General Pinochet and experienced unprecedented changes. The economy was fully privatized. The Central Bank (Federal Reserve) became fully independent. The country withdrew from the Andean Pact, opening the economy to foreign investment without limits or quotas, and even and low tariffs were enforced.

The presidency changed three times during the first few years after Chile regained democracy; but despite the many changes, there has been general agreement on basic government policy between right and left.

DL 701 remains the cornerstone of the incentive policy for encouraging plantations. Eduardo Frei's government opened up DL 701 further, with more incentives for small forest owners and those in remote or fragile areas.

The Native Forest Law has been under discussion for nearly 12 years in the Chamber of Representatives. There is some concern about the illegal conversion of native forest to plantations.

Options for achieving increased sustainability

Environmental Framework Law

An important step in the design and implementation of Chile's environmental policy is Framework Environmental Law No 19,300 (1994), which gave the country environmental legislation and an institutional framework.

The law sets out a system of environmental evaluation that applies to industrial-scale projects in all sectors, not just forest management. New regulations pertaining to environmental impact assessments were passed in 1994. CONAMA is responsible for evaluating individual impact assessments.

The government is committed to generating guidelines for major environmental issues, including water, biodiversity and the renewable natural resources of forestry and fishing.

The main public management efforts so far have focused on environmental pollution problems and its impact on health. Mining has been the main sector considered so far.

Objectives of the Framework Environmental Law The law had three objectives that were designed to increase sustainable private-sector participation, while protecting public interest. The general purpose is to promote the sustainable use of renewable natural resources, and manage, develop and protect it for future generations (see Box 6.1).

Options for regulating forestry

Legal procedures Decree Law 701 monitors afforestation. In order to claim benefits, the owner must apply to CONAF to have the land declared 'preferentially suitable for forestry'. If it is in 'use class VI' or above, it is unquestionably forest land; but otherwise must be inspected and technically assessed.

Afforestation without the subsidy Unsubsidized plantations are treated like any other crop. No special requirements are posed on them, nor is an environmental impact assessment required.

Large companies will plant without subsidy if they can still make a profit. This is particularly the case with eucalyptus plantations in the best sites, close to the mills, in short rotations.

For the average small- or medium-sized forest owner, the subsidy is very attractive, particularly for long rotations; as a result, they will either plant in forest lands or try to reclassify marginal agricultural lands.

Box 6.1 Objectives of Chile's Framework Environmental Law No 19,300 (1994)

Objective 1

Improve the condition of the renewable natural heritage.

Actions:

- Recover degraded areas and resources.
- Prevent deterioration of the natural heritage.
- Protect the natural heritage.
- Develop and research the natural heritage.
- Fulfil international commitments.

Objective 2

Apply and develop management instruments.

Actions:

- Strengthen and coordinate the existing public management system regarding natural heritage.
- Improve the institutional legal framework.
- Generate, disseminate and integrate management information.
- Improve instruments for sustainability and develop new ones.
- Promote public–private cooperation.

Objective 3

Create the necessary conditions for the sustainable use of the natural heritage.

Actions:

- Promote the participation and involvement of the public in management.
- Promote environmental education and a culture of caring for the natural heritage.
- Create opportunities for indigenous, fishing and rural communities.

CONAMA has all of the legal tools to enforce the framework.

Contractual procedures If the land is sold, the new owners must still observe the laws governing property preferentially suited for forestry. Their harvest plans must be approved by CONAF and they face stiff fines for not complying with reforestation responsibilities.

Lessons from Chile's experience

Forest certification

There are two initiatives to develop forest certification standards in Chile. An FSC working group is developing standards for native forests and plantations under the name *Iniciativa de Certificación Forestal Independiente* (ICEFI, Independent Initiative in Forest Certification) and based on the generic FCS standards.

A group of private Chilean institutions has also developed the CertforChile Standard. It has already been benchmarked against the FSC standards and has performed well. Both standards address almost the same issues; the main difference is that FSC does not allow genetically modified (GM) trees in certified forests. The CertforChile standard enforces a precautionary principle: in general terms, it recognizes that the large-scale application of new technologies, species or varieties will be recognized only if a prior impact assessment has shown that the social, environmental and economic impacts are acceptable. The nine principles of the CertforChile standard are set out in Box 6.2.

Both large and small forestry companies in Chile have supported the idea of developing a national standard that would be recognized by other certification initiatives, including FSC and particularly the Pan-European Forest Certification Framework. FSC has certified 5000ha or 0.08 per cent of native forests and 247,000ha or 12 per cent of plantations.

Attitudes towards certification have matured. Today, forest owners accept and are willing to undergo performance-based certification. Customers, particularly wood-chip customers in Japan, are pressing for certification. All of the largest companies are already certified ISO 14000. In general, the FSC has succeeded in pushing its logo only in foreign-owned forests; in general, Chilean companies and the government favour a national certification system recognized under international agreements.

On the whole, the government has a positive attitude towards certification as a means of securing markets. A Forest Ethics campaign to boycott Chilean exports of radiata pine unless they bear the FSC logo has aroused resistance from the public, the companies and the government.

According to a CONAF official, CONAF is working towards applying the Montreal Process criteria and indicators in monitoring and controlling management standards in plantations.

Pilot trials carried out by CertforChile proved that as a result of preparing for ISO 14000 certification, most company's forest management standards are very high.

Conclusions and ways forward

The state seeks a balance between the rights of individuals and society and their obligations to maintain Chile's natural heritage. The state must ensure

Box 6.2 The nine principles of the CertforChile standard

With a view to providing a better understanding of some of the key issues for proper forest and plantation management in Chile, selected criteria from the CertforChile standard are presented, alongside their respective principles below.

Principle I

The use of forest resources must be planned and managed in order to provide a sustained flow of products and services in successive rotations. Management must be in accordance with a long-term comprehensive management plan for the forest management unit (an FMU can belong to a single owner or to a group) appropriate to the scale of the operations. The management plan must be prepared before operations are started.

Criterion 1.1
Forest managers are formally committed to sustainable forest management and can demonstrate their intention to continue with forestry activities in the FMU for at least one more rotation.

Criterion 1.2
There is a forest management plan appropriate to the scale of operations, which is implemented. The plan clearly specifies the objectives of forest management.

Criterion 1.6
The large-scale application of new technologies, species or varieties is adopted under a precautionary principle. These are only adopted if a prior impact assessment has shown that social, environmental and economic impacts are acceptable.

Principle 2

The use of forest resources must be planned and managed in such a way that the environmental value of native ecosystems contained in the FMU are protected and that negative impacts on biodiversity are minimized.

Criterion 2.1
Plantations will not be established in areas that contain forests or other native vegetation types of high conservation value or in commercially productive native forests.

Criterion 2.2
The planning system of the FMU takes into account the existence, environmental value and management needs of different types of native vegetation.

Criterion 2.3

Areas with high conservation value native vegetation are managed in order to maintain the biodiversity that they provide.

Principle 3

Forest resources should be managed in order to maintain their health, vitality and productivity by protecting them from fires and other damaging agents.

Criterion 3.2

The control of pathogens and other damaging agents is carried out according to the principles of integrated pest management. Activities are carried out so as to minimize negative environmental impacts. The entire forest resource is protected from damage by animals.

Criterion 3.4

The control of plantation weeds should be carried out in order to maximize the growth of trees while minimizing the use of herbicides.

Principle 4

Forest resources are managed so as to promote soil conservation and to minimize adverse impacts on the quantity and quality of water resources, taking particular account of the needs of downstream communities.

Criterion 4.1

There is a classification of soil types according to their fragility or level of erosion and of water bodies and rivers appropriate to the scale and intensity of operations.

Criterion 4.3

Roads are planned, built and maintained so as to minimize erosion and the carriage of sediments into watercourses.

Criterion 4.5

The planning of forest management is done so as to avoid the contamination of water, as well as the reduction of water flow in areas with water shortage or where rivers supply downstream communities.

Principle 5

Forest managers must respect the traditional and customary uses and rights of local communities, maintaining good neighbourly relations with them and supporting the development of local capacities which contribute to the improvement of their quality of life.

Criterion 5.1
Forest managers have knowledge of the impact of their activities on local communities.

Criterion 5.2
Forest managers make contributions towards improving the quality of life of surrounding communities.

Principle 6

Forest managers will take into account declared agreements and documented commitments and will respect the legally established rights and the traditional knowledge of indigenous peoples to use and manage their lands and resources.

Criterion 6.1
Forest managers are aware of the presence of indigenous peoples in the area of their management activities. They know the rights of these peoples and respect them.

Criterion 6.3
Indigenous communities are fairly compensated for any use of their traditional knowledge about forest management and the specific use of indigenous plant species by forest managers.

Principle 7

Forest managers will respect the rights of the forest workers, compensating them fairly and equitably, safeguarding their health and safety at work.

Criterion 7.1
Forest managers ensure that forest workers are trained so that they can carry out their work in a productive manner and that they also have opportunities for development.

Criterion 7.2
Forest managers respect the rights of workers to the benefits of organizing themselves and of collective bargaining.

Criterion 7.5
Forest managers provide fieldworkers with adequate transport, accommodation, rest and food.

Principle 8

Forest managers respect the laws of Chile and international agreements and legally binding treaties, and will take into consideration any other agreements and treaties to which Chile is a signatory.

Criterion 8.1
Forest managers know and respect national legislation applicable to their activities.

Criterion 8.2
Forest managers know and respect legally binding international treaties to which Chile is a signatory.

Principle 9

Regular monitoring of the forest resources, the management system and the responsible companies and owners of the FMU will be conducted with the purpose of evaluating the progress in achieving the stated principles.

Criterion 9.1
There are procedures for regular evaluation of the condition of the forest resources and of the significant environmental social and economic impacts of forestry operations. Monitoring procedures are consistent, replicable over time and allow for comparison of results of change.

Criterion 9.3
There is a procedure to trace and account for the quantity of wood coming from certified forests (the FMU itself or purchased from other certified forests) that is sold to processing plants, from its origin in the forest to its point of sale (a procedure known as chain of custody).

and promote the responsible management of ecosystems and natural resources, the maintenance of a supply of goods and services, and the generation of economic and social opportunities for the long term.

The main challenge is to restore degraded soils. Soils must maintain their productive capacity and the vitality of their ecosystems. The sustainable management of soils must be integrated with management of the forests, water and biodiversity.

A benchmark in Chile's forestry history was the passing of DL 701 in 1974. Thereafter, the planted area grew at exceptional rates, all in the hands of private owners; new industries also began their operations or new installations were created, forest products were diversified, and forest exports expanded phenomenally.

Industrialization

Chile's plantations comprise two main species, the long-fibre radiata pine and the short-fibre eucalyptus. During the last decade, more than 80 per cent of timber came from radiata pine plantations.

The introduction of species such as radiata pine and eucalyptus has led to the long-term survival of the forestry sector. Numerous new industries were created, which in turn created additional demand for timber from radiata pine and eucalyptus plantations.

This led to the 'industrialization' of Chile's forests and a rapid growth of the nation's forestry-based industry. Chile's pulp and paper industry began in 1965 and further expanded in 1972 with the establishment of the Arauco and Constitución mills, which were followed in the 1980s by at least three additional pulp mills and the expansion of others, and the installation of several very modern and large saw mills and, lately, by a veneer mill. During the early 1990s, related remanufacturing industries were established, completing this success story of privately owned plantations providing a large amount of raw forest material.

The dynamic growth of modern, forestry-based industries has also promoted employment. Direct employment may not have increased in direct relation to output, particularly in the modern pulp and paper and saw-milling sectors; but the forestry sector is still characterized by production pluralism with many very large, medium-sized and small manufacturing mills generating most of the employment. Indirect employment has been more than triple direct employment, with a very high 1998 record of about 124,000 employees, down to the current 117,000. Of this, 12 per cent are in related services, 54 per cent in the industry and 34 per cent in logging and silvicultural operations.

Economic impacts

Forestry added regional balance to the notoriously regionally unbalanced economic fabric of Chile. Jobs were created for skilled labour, managers and professionals, along with the saw mill workers earning lower wages. The numerous low-income self-employed foresters and the many seasonal and frequently very poor workers engaged in timber extraction and small saw mills also benefited from the increased demand for labour.

The success of radiata pine plantations greatly strengthened the forestry sector's capital formation linkages with the rest of Chile's economy: it increased the rate of accumulation of the total forest capital stock; facilitated the establishment and expansion of industries that depend upon using plantation forest capital, such as pulp, paper, veneers and so forth; generated demand for previously arid or uncultivated land that was suitable for afforestation; and generated demand for machinery and equipment to be used in extraction and transformation of timber, and in providing ancillary services (much of this machinery and equipment was imported).

Using forest resources facilitated capital formation in housing, which for decades was Chile's most important area of investment. Exploiting Chile's native and plantation forest capital permitted import substitution and export promotion in forestry-based industries and the transfer of technological capital related to this sector from abroad to Chile. The high rate of native forest depletion due to fire and insects prompted technological transfer of fire protection and control services to Chile from Germany and Canada. Strong competition among suppliers and an artificially low foreign exchange rate have kept capital goods prices low, and have led to a higher degree of mechanization than would have been achieved had the foreign exchange rate not been so overvalued.

From Centrally Planned Economy to Vigorous Rural Enterprise: China

Jintao Xu and William Hyde

Introduction

Twenty-five years of rapid economic growth have created great change in China's demands on its forest and considerable change in the forest itself. Its experience is instructive for any country undergoing the transition from a poor agrarian economy to a diversified and developed economy, and particularly for any country contemplating or currently experiencing the transition from centrally planned and publicly owned forestry operations to an expanded role for markets and private initiative.

China began its remarkable transition with the introduction of rural reforms in 1978. The summary effects are well known: average annual growth in gross domestic product (GDP) regularly exceeding 10 per cent and more than a quadrupling of per capita income by the year 2000. The rural population was the first to benefit, and rural incomes increased more than sixfold over this period. As industrial, financial and then trade reforms followed the rural reforms, nearly all sectors of the economy and all members of Chinese society benefited. The market now plays a greater role in both private enterprise and the remaining state-owned enterprises. Nevertheless, many activities are still regulated and regulations governing private forestry are particularly burdensome. Meanwhile, the public sector of China's economy remains substantial and many state-owned enterprises continue to absorb public financial support far in excess of their contribution to the economy. Reform of state-owned enterprises is the focus of government policy as China enters the 21st century, and the performance of the state-owned forest enterprises is an important feature in this context.

Basic statistics and future outlook

The People's Republic of China is the world's largest country, with a population of 1.26 billion in 1999, about 22 per cent of the global total. Its land area of 9.6 million square kilometres ranks third in the world, after Russia and Canada (FAO/UN, 2001).

China's climate varies from tropical in the south-east to cool temperate in the north and its annual precipitation ranges from 2000mm in the south-east to 100–200mm in the arid north-west. Its topography varies from mountains and high plateaus in the west to basins and plains in the east.

China's economy has grown rapidly over the last quarter century. Per capita GDP has quadrupled at a real annual rate of approximately 8 per cent, lifting some 300 million people out of poverty. Farmers and the general rural population have benefited most (see Table 7.1). Rural economic growth is the

Table 7.1 *China's economic growth, 1978–1998*

	1978ᵃ	1998	% Change
GDP (1978 monetary value)			
Aggregate	362	2312	538
GDP per capita	379	1869	390
Rural GDP per capita	133	945	610
Forest products			
Production (thousand cubic metres)			
Logs	51,673	55,557	7
Lumber	11,055	17,876	62
Wood-based panels	1017	10,563	939
Paper and paperboard	4390	21,256	385
Imports (thousand cubic metres)			
Logs	1870	4820	158
Lumber	75	1679	2139
Wood-based panels	258	1977ᵇ	666
Paper and paperboard	767	5760	651
Exports (thousand cubic metres)			
Logs	28	63ᵇ	125
Lumber	13	255	1862
Wood-based panels	10	598	5980
Paper and paperboard	229	250	9
Forest cover			
Area (million hectares)	115	154	35
Volume (million cubic metres)	8801	10,086	15

Notes:
a Imports for 1981, exports for 1983.
b 1997 data.

Source: China Statistics Yearbook, 2000, in SFA, 2000

most visible effect of the government's economic reform programme which, beginning in 1978, promoted economic efficiency by introducing greater opportunity for private activity and expanding the role of markets in resource allocation. Private enterprise now accounts for about 33 per cent of GDP, a share that is almost comparable to the 37 per cent contributed by state-owned enterprise (Kunge, 2000).

The role of the forest in China's economy

China's land area is allocated as cropland (10.4 per cent), permanent pasture (42.9 per cent), forest and woodland, including grassland and some unstocked wasteland (16.6 per cent), and other (30.1 per cent). The total area in forest is 154 million hectares. While this aggregate forest area is large, the fifth largest of any country in the world, it is small relative to China's population and relative to the economic demands placed on it. Per capita forest area is only 0.1ha, considerably less than the global average of 0.6ha per capita (FAO/UN, 2001).

The growth of the forest sector has been more variable and, in some parts, not as rapid as growth in the general economy (see Table 7.1). Log production increased only slightly since 1978, while lumber production has increased a moderate 60 per cent in 23 years, a rate of approximately 2 per cent annually. The paper industry, on the other hand, has kept pace with double-digit annual growth in the general manufacturing sector, expanding to almost five times its level in 1978. China's forests have not been able to keep pace with the expansion in manufactured wood products and imports have provided the difference. China is now the world's second largest importer of timber, and forest products, in general, are China's largest import commodity group (SFA, 2002b).

Forestry accounts for only about 1 per cent of China's GDP; but the forests are an essential source of energy for 40 per cent of the rural population and they supply virtually all of the wood used by the construction industry. China's forests will undoubtedly continue to provide wood as a raw material; but the relative importance of this to the economy is uncertain. The rapid growth of the paper industry is expected to continue as the country continues to develop and the demand for paper products tends to be income elastic.[1] Other wood products tend to be less income elastic, and the demand for fuelwood may even decline for households above some level of income. The demands for these other wood products may be less imposing on China's forests over time.

The demand for the environmental services provided by forests will probably grow with time for at least three reasons. Firstly, 38 per cent of the country is affected by soil erosion and 27 per cent is desert, and this desert area is expanding by 250,000ha annually (SFA, 2002b). We can anticipate that both government projects and individual farmers will continue to plant trees to help control this problem (Zhang 2002; Yin and Hyde, 2000). Secondly, the government embarked on the massive Western Regional Development Programme (WRDP) in 2000 that includes major components to protect existing forests and add substantially to reforestation through the year 2010. Thirdly, as economic development proceeds and household incomes continue to grow, the demand for forest-based leisure activities will also increase.

The forest resource

Figure 7.1 outlines China's regions and the provinces within them. China organizes its official forests statistics by region and administrative responsibility. Table 7.2 identifies the land area under state-owned and collectively administered forest cover recorded in each of the six principal geographic regions in the three inventories from 1977 to 1998.[2] The largest forests are in the north-east, the south-west and the south central regions.

Forest administration

China broadly organizes the administrative responsibility for its forests into two categories: collective forests and state-owned forests.

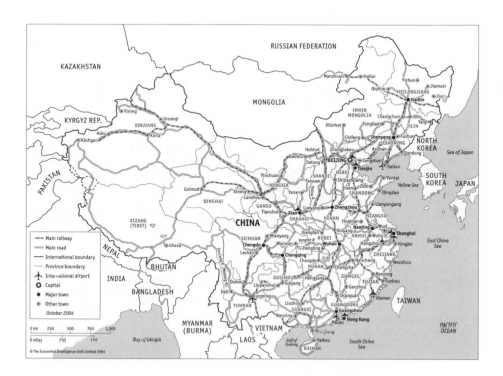

Source: © Reproduced by permission of the Economist Intelligence Unit

Figure 7.1 *China and its provinces*

Table 7.2 *China's forest land area^a*

Region	1977–1981	1984–1988	1994–1998	Percentage change
North-east				
State owned	21.37	20.55	23.50	9.97
Collective	2.33	3.96	4.20	80.26
Total	23.70	24.51	27.70	16.88
North-west				
State owned	2.48	5.30	5.78	133.06
Collective	1.96	2.80	3.19	62.76
Total	7.24 (4.44)^b	8.10	8.97	23.90 (102.03)^b
South-east				
State owned	1.32	1.73	2.49	88.64
Collective	10.84	12.01	16.55	52.68
Total	12.16	13.74	19.04	56.58
South-west				
State owned	n.a.	8.15	9.42	
Collective	n.a.	12.24	17.39	
Total	23.53	20.39	26.81	13.94
South Central				
State owned	1.78	2.18	2.55	43.26
Collective	10.32	14.92	22.44	117.44
Total	17.35 (12.10)^b	17.10	24.99	44.03 (106.53)^b
North China Plains				
State owned	13.26	13.11	14.18	6.94
Collective	1.55	2.13	3.43	121.29
Total	14.83 (14.81)^b	15.23	17.61	18.75 (18.91)^b
All China				
State owned	n.a.	51.01	62.01 (57.93)^c	
Collective	n.a.	48.06	67.19	
Total	98.81	99.07	129.20 (125.12)^c	30.76 (26.63)^c

Notes:

a China's forest survey divides 'forests' into 'economic forests', 'bamboo forests' and 'forest stands'. Economic forests are tree crops such as orchards and rubber. Forest stands are what most of the rest of the world calls forests, and are what is reported in this table.

b In China's second inventory (1977–1981), the data from some smaller provinces was included in the total but not in the disaggregation into state-owned and collective lands. The first entry in our 'total' row for three regions is the total reported to the national inventory. The bracketed entry is the sum of recorded state-owned and collective lands for the larger provinces in those three regions.

c The bracketed entries are the sums of our regional data. The larger entry is the official national aggregate.

Source: SFA (Forestry Statistical Yearbook), various issues

Collective forests

Collective forests are those that were managed by local forest and agricultural collectives until 1978. Since then, individual households have gradually assumed long-term contractual responsibility for more and more of the collective lands. The area under collective forest administration expanded from less than 48 million hectares in 1978 to over 67 million hectares by the mid 1990s. By the time of China's fifth official forest inventory during the mid 1990s, the collective forests accounted for about 60 per cent of China's total forest land, and individual households managed approximately 80 per cent of that (SFA, 1984–1998).

Collective forest management still has its problems. For example, taxes are much higher on forest products than on the products of competing land uses (Liu et al, 2002; Ruiz Perez et al, 2002), and government procurement restrictions still limit private timber opportunities in most regions of the country. Nevertheless, the performance of China's state-owned forests is altogether more problematic, and this latter component of China's forests has become the greater focus of current forest policy.

State-owned forests

The state-owned forests include about 135 state forest bureaus and approximately 3000 independently operating forest farms. The bureaus were originally set up as logging operations for natural forests and control their own forest farms. Most also control their own wood processing facilities. Many of the independent forest farms were originally set up to manage forest plantations.

The reporting hierarchy for both institutions has changed several times. Both currently report to local or provincial governments for personnel and planning but continue to receive financial assistance from the central government and are subject to the forest management guidelines of the State Forest Administration (SFA) (the Ministry of Forestry before 1998).

The forest area under state administration has expanded from approximately 51 million hectares in 1978 to 62 million hectares by the mid 1990s. Government investment in the state-owned forests expanded similarly over this period – at a 7.9 per cent annual rate. Yet, 80 per cent of the state-owned forest enterprises (SOFEs) had exhausted their mature timber and more than one half of them operated in financial arrears by the mid 1990s (Zhang, 2001; Zhang, 2002). Meanwhile, the demands on these state-owned forests for wood as a raw material, for fuel and for environmental and recreational services have continued to grow.

State versus collective forests

The land areas in both state-owned and collective forests have expanded gradually; but the land area in collective forests is growing more rapidly in most regions. Nevertheless, the standing forest volume per hectare remains almost three times greater on the state-owned forests in 1998 than the collective forests. This is partly due to the fact that some state-owned forests are inaccessible and biologically mature, while all collective forests are accessible. Therefore, the collective forests are harvested at younger ages and smaller volumes, while some state-owned forests are not harvested at all.

Other reasons for such a large difference may be related to the difficulties encountered in measuring the newest young growth during forest inventories;[3] or it may be due to China's accounting system for forests;[4] or it may be due to a range of factors that discourage private investment. It will be interesting to observe whether the standing volumes in China's forests increase rapidly as the effects of reform incentives for planting and management become absorbed into the measurable growth that will be captured in future forest inventories.

The collective forests contain nearly three times the total area and nearly twice the total volume in plantations as the state forests, plus most of the economic and bamboo forests. These observations are not surprising – for reasons that are similar to the differences in inventory volumes just discussed. The state forests contain most of the remaining inaccessible natural forest, including some managed stands and plantations, while most of the collective forests are managed and plantations account for a larger share of them. The collectives and individual households manage the larger shares of economic forests (generally, horticulture) and bamboo because the products of these forests are not within the normal responsibility of state forests.

Historical overview

Forest nationalization and collectivization: 1950

China nationalized all forestry enterprises and confiscated feudal forest land in the mountainous areas for redistribution to farmers in 1950, shortly after the Communist government assumed power. In effect, two systems of ownership were established: state-owned forest land under the management of SOFE and individually owned mountain forest lands.

A new national policy of collectivization began in 1953 and the era of co-operatives and people's communes in mountainous and forest areas began in 1958. Private forests were eliminated, and the titles to mountainous forest land became less clear. Productivity declined. Subsequent periodic policy revision and improved management incentives failed to have permanent beneficial impacts. A quarter century of poor economic performance and a decline in the forest base followed.

Transfer of collective forest lands to households: 1978–1984[5]

The period of reform (1978–1984) was characterized by decollectivization and a relaxation of the Unified Procurement Pricing System (UPPS) for agricultural products whereby agricultural households regained land use rights – a policy known as the Household Responsibility System (HRS) – and were eventually able to sell most of their production at market prices.

Subsequently, households gained land-use rights to collective forest lands, as well,[6] under a comparable policy known as the Contract Responsibility System (CRS). Change was rapid and by 1984, 60 per cent or 30 million hectares of collective forest land had been transferred to 57 million households. Various forms of non-state forest enterprises (rural forest co-operatives, share-holding groups, joint venture firms, private forest farms, etc) have emerged since then from this household-based forest system.

The SOFEs were largely unchanged during this initial period of reform, although some of them did return land to the collectives and to households as the ownership rights and boundaries of forest properties were re-established.

The central government maintained an active role in afforestation and reforestation during this period; but the role of individual households increased. Some households drew on their own resources to reforest (e.g. Yin and Newman, 1997; Liu and Edmunds, 2002). The central government also invested in programmes such as the Three North Forest Protection project (otherwise known as the Green Great Wall) for erosion control and in the silvicultural activities of the SOFEs. However, the rural population was no longer a passive participant. When local authorities looked for land upon which to plant trees, they had to negotiate with the landholders – the local farmers – and the farmers had the right to demand compensation. Overall, as indicated in Table 7.2, the area of collective forest lands nearly doubled, while the area under SOFEs remained stable between 1977–1981 and 1984–1988 (the second and third national forest inventories, respectively).

Industrial and financial reform: 1985–2000

Subsequent reforms focused on industrial and financial sectors in the urban areas and development of non-agricultural enterprises in the rural areas. A Manager Responsibility System for state-owned enterprises (comparable to the HRS and CRS) permitted managerial discretion over the use of variable inputs. Then, during the 1990s, the government permitted the sale of some state-owned enterprises (SOEs) and permitted others to go out of business. In 1997, it permitted managers of SOEs to release redundant labour – and 7 million workers were absorbed into the non-state economy (Hyde et al, 2002b).

In the rural areas, smaller non-state enterprises known as township and village enterprises (TVEs) developed. These became the fastest growing component of China's economy. Ninety per cent of China's paper mills are now TVEs, and these have grown more rapidly than the state-owned paper mills. Both saw mills and paper mills more than doubled their production between 1978 and 1989. The paper mills continue to expand their production to this day (Xu, 1999).

Four factors had important effects on forestry during this period: auctions of wasteland; the liberalization of the UPPS for timber; general economic growth; and the liberalization of foreign trade. The government began auctioning barren forest lands (the 'four wastelands') for afforestation in

1993 and allowed private operators to compete in these auctions. By 1996, the management of 3.7 million hectares had passed into private hands via this method. The practice of selling forest land through public auctions has now extended to lands with juvenile and mature stands of timber.

The UPPS has been gradually relaxed, with the government now accounting for less than 10 per cent of timber purchases (Zhang et al, 1994; Waggener, 1998). Nevertheless, government regulations on timber harvest levels and shipments remain strong and timber markets remain underdeveloped in some regions.

General economic growth, rather than any specialized forest policy, was responsible for growth in forest management during this period. Paper-making is a good example of this. Paper production grew at a 13 per cent annual rate in China after 1984, a rate in excess of the 8–10 per cent rate of annual growth in GDP. As a result, the industry's demand for wood fibre grew and created a price incentive for expanding forest management. More generally, Zhang et al (2000) determined that a 1 per cent increase in per capita GDP explained a 0.59 per cent increase in plantation forest area in one province (Hainan). Rozelle et al (2002) determined that a 1 per cent increase in the light industry share in the full economy of China was responsible for a 0.13 per cent increase in forest land.

On the other hand, trade liberalization absorbed some of the increasing demand for woody raw material, and new restrictions on logging in 1998 ensured that log imports would become even more important. As a consequence, log imports nearly tripled (from 4.8 million cubic metres to 13.6 million cubic metres) between 1999 and 2001 (China Customs Office, 1999–2001).

On aggregate, the area under collective forest lands increased by 40 per cent to 67 million hectares between 1984–1988 and 1994–1998 (third and fifth national forest inventories). The forest area under state lands, managed by the SOFEs increased by 20 per cent to 62 million hectares over the same period – as can be seen in Table 7.2. All of the increase was newly planted and can be considered forest plantation, although in reality the total area in new plantations was even larger as additional plantations replaced some natural forests.

Plantation history

China has the largest area in forest plantations of any country: 46.7 million hectares, or approximately one quarter of all plantations in the world (SFA, 2000a; FAO/UN, 2001). These plantations comprise 22.5 per cent of China's forest stands and 30.4 per cent of the country's total forest land, including economic and bamboo forests. They contain a smaller share of the country's total standing forest volume (9 per cent), partly because most of these plantations are younger than the remaining natural forest. Their younger trees are smaller;[7] but they are also growing more rapidly than the trees in the natural forest.

The total area successfully planted since 1949 has been approximately 100 million hectares. A little more than 50 million hectares have been planted, then eventually harvested or converted to other land uses, without replanting. Almost 50 million hectares remain under plantations today.

Plantations were harvested at an increased rate during the early and mid 1990s (between the fourth and fifth national forest inventories) due to economic growth and the increased demand for woody raw material. However, afforestation in new plantations outpaced the harvesting of existing plantations and the total area in plantations rose 12.42 million hectares during this same period (see Table 7.3).

Table 7.3 *Change in China's plantation area*

	Area of closed plantation	Net increase between inventories
First inventory, 1973–1976	23.69	n.a.
Second inventory, 1977–1981	22.19	−1.50
Third inventory, 1984–1988	31.01	8.82
Fourth inventory, 1989–1993	34.25	3.24
Fifth inventory, 1994–1998	46.67	12.42

Note:
1 Between the period of the fourth and fifth inventories, an area of 1.75 million hectares of plantation forests were degraded to open forests, shrubs or barren land, and 1.66 million hectares were converted for other types of land use. The estimated loss in plantation area was 6.69 million hectares, 1.34 million hectares per year. This was more than offset by 19.01 million hectares of new plantation.

Source: SFA (2000a)

Table 7.4 traces the pattern of plantation growth for the collective and the state-owned forests from the third to fifth national forest inventories. It is clear that the area under plantation is growing rapidly and that the collective plantations contribute a large (greater than 70 per cent) and growing share due to the policies of the first 15 years of reform, which had their greatest effects on private and household incentives. Households themselves were responsible for 50 per cent of the investment in plantations by 1988 (Li et al, 1988), and they probably have been responsible for an even larger share since then.

The government has targeted a further 10 per cent increase in total land area devoted to forests, an increase from 16.6 per cent in the fifth national forest inventory to 26 per cent by the year 2050 (SEPA, 1999). This means that plantation area will continue to grow. Furthermore, the government's 1998 ban on logging in natural forests makes plantation development all the more urgent – in order to serve the demands for raw materials that can no

Table 7.4 *China's plantation development and growth*

Gross plantation area (million hectares)	Total	State	Collective	Share of collectives
1984–1988	18.74	5.48	13.26	0.71
1989–1993	21.37	6.24	15.13	0.71
1994–1998	29.14	7.70	21.44	0.74
Volume of plantation (million cubic metres)				
1984–1988	529.85	213.20	316.65	0.60
1989–1993	711.98	295.27	416.71	0.59
1994–1998	1012.99	378.33	634.66	0.63

Source: SFA, 2000a

longer be supplied by the natural forest. It is anticipated that the collective sector (mostly household managed) and the more market-oriented SOFEs will be responsible for the largest shares of these new plantations, responding, respectively, to market incentives and to new management incentives given to managers during the late 1990s.

Development and status of the collective forests

Shortly after agricultural households began receiving HRS rights to agricultural land in 1978, they began showing an interest in similar rights to the small remaining forest lands of the agricultural collectives. At the same time, households in the forest collectives began showing interest in similar rights to the lands of these collectives. As had been the case with agriculture, the local authorities recognized the administrative advantages of the land transfers, and households recognized the advantages of personal choice and eventually increased productivity.

Transferral of collective forest lands to households

The collectives transferred more than 31 million hectares of forest lands to 57 million households by 1984. The households themselves began their own afforestation and reforestation activities, increasing the forested area of the collectives by 33 per cent by 1988 (the end of the third forest survey) and by 86 per cent by 1998 (the end of the fifth survey). By 1998, households managed 52 per cent of China's total forest area, incorporating 74 per cent of China's forest plantations, 92 per cent of its economic forests and 93 per cent of its bamboo forests.

Method of land distribution

The distribution of forested land was egalitarian. Forest land was first divided into plots according to species, age, density, site and soil quality, and distance from the village and then distributed in such a way that households of similar family size received similar mixes of high and low quality forest, accessible and less accessible forest and so forth. As a consequence, land on one slope was often distributed among many households in one village. The average household in southern and south-western China obtained four to five plots totalling just 1ha to 2ha.

The objective of equity among households was achieved; but the many fragmented and scattered plots increased the demands on the managerial resources of each household as small plot size increased the difficulty and reduced the cost effectiveness of boundary marking, protection[8] and most other forest operations. The problem of highly fragmented holdings was compounded by unclear boundaries, which were also a consequence of the land redistribution process and remain a source of conflicts today.

The reforms, as they were applied in China's collectives, were pragmatic and variable. They were adapted to local conditions and they varied with local interests, local authorities and, over time, with the changing economy. In some places management problems associated with small forest plots were addressed by households or the collective authorities through land transfers and plot consolidation. Occasionally, the pattern of fragmented holdings led to the development of shareholding agreements.

After bouts of generalized inflation during the mid 1980s, some local authorities expressed doubts over household management and revoked some household contracts on forested land, re-imposed the restriction of government procurement and increased taxes. However, many of the newer restrictions were gradually reduced or eliminated and when the earliest contracts were renewed during the middle 1990s, the households responded with increased confidence in their ability to invest and recover the fruits of their investments from their forested lands.

Lessons

Two crucial lessons have emerged from these experiences:

1 When the economic conditions are right, farmers will make long-term investments, such as in forest management, but will also invest even when there is no direct financial advantage.
2 An uncertain policy environment makes farmers reluctant to invest in forest management.

Both of these points have been demonstrated by Yin (Yin and Newman, 1997, Yin and Hyde, 2000), who observed that forest cover expanded from about

5 per cent in 1977 to 11 per cent by 1988 in the agricultural northern plains, mainly as a means of protecting against the region's severe problem with wind erosion. Yin used an agricultural production function to assess the effects of improved property rights (HRS), market liberalization and the conservation investment in trees, as well as the standard agricultural inputs, on agricultural output. He concluded that improved property rights provided the stimulus for approximately 10 per cent of the eightfold increase in agricultural output between 1978 and 1984. Investments in trees explained 5 per cent of the increase in agricultural output through 1984 and 20 per cent thereafter. In effect, private management provided a level of forest protection against the non-market effects of soil erosion in China's northern plains.

Yin contrasted the experience of the northern plains (where agriculture was so important that the authorities paid little attention to forests) with that of a part of the Southern Collective Forest Region, just south of the northern plains, a region where forestry was traditionally of greater importance. The south was the initiating region for both agricultural and forest reforms and it contains about 85 per cent of China's collective forest lands. Land and tree tenure had changed three times over 20 years. Consequently, southern households were cautious of new contracts for forest land use. Their doubts were confirmed when the authorities in some parts of the south rescinded some of those contracts. Yin showed that farmers in both regions responded rationally. As market prices rose in the north, farmers first harvested the trees on their new lands; then, as prices continued to rise, they planted more trees – and harvest levels continued to increase over time as a result. As prices rose in the south, farmers harvested their trees; but as prices continued to rise, they did not invest. They were concerned that they would lose the benefits of their investments as a result of future policy revisions. Consequently, harvests in the south declined over time. Policy has been more consistent since the mid 1980s and southern households have responded accordingly. Every official forest survey since the late 1980s shows an increase in both the standing forest volume and the area of household managed forest.

Income distribution

Have the successes of household forest management in China improved household income? How have they affected the poorest rural households? Questions about forestry and the rural poor arise because extensive forest cover often occurs together with poverty and isolated human communities. We are unable to estimate the quantitative effect of growth in the forest sector on China's rural households, although we do know that the incomes of rural households, in general, increased more than sixfold since the beginning of reforms in 1978; undoubtedly, some of this increase came from forestry.

There is evidence to address these questions, however, from the bamboo sector. Bamboo is a sub-sector of forestry in China, and bamboo and commercial timber are similar in their productive inputs. Both use rural labour and land and neither is a significant competitor for agricultural capital. Bamboo also competes in many of the same product markets as timber, providing raw material substitutes for timber in paper and various construction activities. Bamboo production has grown more rapidly than commercial wood – from 3.2 million hectares and 4.4 million metric tonnes of output in 1980 to 4.3 million hectares and 14.2 million metric tonnes in 1999 (Ruiz Perez et al, 2002). Its more rapid growth is suggestive of the development pattern that would have occurred for commercial timber under more liberal market conditions.[9] This contention was reinforced when China introduced a logging ban over large areas of the country in 1998. Bamboo prices immediately rose 5–10 per cent and the land area in bamboo plantations grew nearly 17 per cent in one year (CFIC, 1998).

Ruiz Perez et al (2002) reviewed the bamboo experience for six counties along an east–west gradient from the Anji Province on the Pacific Coast to more remote inland parts of Hunan and Sichuan provinces. They observed that improving bamboo markets provided an alternative income source for many farm households and an opportunity for some households to diversify and specialize. They divided farm households by income group and observed that households in all income groups gained from rising bamboo prices and market opportunities; but middle and higher income groups gained relatively more. Bamboo opportunities were important; but off-farm employment opportunities were even more important sources of increasing household income, and middle and higher income households obtained more of these opportunities, as well. Bamboo offered a supplementary income opportunity for many households, but many poorer households remained poor and their gains from bamboo development were small.

We can hypothesize that commercial forestry will follow a similar pattern if China further liberalizes its forestry markets. Farm households will take advantage of the opportunity. They will supplement their incomes and diversify their production, and some will eventually specialize in particular forest products. The greatest opportunities will occur where market access is good, and China's efforts in the 1990s to improve rural roads and other infrastructure will be important in this respect. However, we can anticipate that the experience of China's poorest bamboo farmers will be duplicated, as well. The incomes of poor forest households will improve; but the poorest households are unlikely to be the primary beneficiaries of forest-based development and many will remain poor even after forest development.

Development of the state-owned forests

The development and performance of state-owned forest plantations cannot be traced directly as the data is not readily available. However, we can identify the most important developments in the overall investment and management of the state-owned forestry enterprises.

Investment

Since 1978, the central government's funding has accounted for less than half of all investment in SOFEs (including investment in logging and wood processing), but more than half of the investment in SOFE silvicultural activities (see Figure 7.2). The differences are composed of provincial and local government expenditures and investments from the retained earnings of the SOFEs. The central government's share of all investment in the forest industry has declined with industrial reform and the 'hardening' of budgets for all state-owned enterprises. In contrast, the government's share of silvicultural investments has increased, especially since 1997, with its growing recognition of the importance of the environmental services provided by forests.

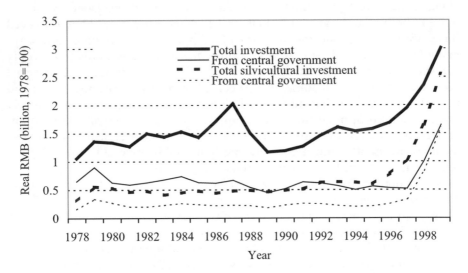

Source: China Statistical Yearbook 2000

Figure 7.2 *State investment in China's state-owned forest enterprises*

Total government investment in SOFEs increased at a real average annual rate of 3.9 per cent between 1979 and 1997. This is an aggregate rate that reflects a 7.9 per cent average annual rate of increase for silvicultural investments, but only a 2.3 per cent average annual rate for investments in logging, wood processing and other activities. This pattern has since shifted, with recent decisions to leave the forest industry to the private sector and to improve the environment. During 1998 and 1999, the government's investment in silvicultural increased sharply, while its investment in logging, wood processing and other SOFE activities actually declined.

These aggregate measures mask the substantial impacts of a few large investments (see Table 7.5). The largest increases in state silvicultural investment

Table 7.5 *Overview of China's main afforestation programmes since the 1970s*

Name of programme	Years	Coverage (area)	Targets	Achievements to date
National Greening Campaign: the National Compulsory Tree-planting Campaign	1987–current			1987–1997, 27.9 billion trees planted
Three North Forest Protection Programme	1978–2050	551 counties in 13 provinces; 40.6 million ha (50% northern China)	Afforestation of 35.08 million ha by 2050	25.67 million ha planted by 1999
Shelterbelt Development Programme along the Upper and Middle Reaches of the Yangtze River	1989–2000	271 counties in 12 provinces	Afforestation of 67.05 million ha	1989–1999, 4.8 million ha planted
Coastal Shelterbelt Development Programme	1991–2000	195 counties in 11 provinces	Afforestation of 3.56 million ha	1991–1999, 1.08 million ha planted
Farmland Shelterbelt Development Programme in Plain Areas	1988–2000	918 counties in 26 provinces	Set standard	1988–1999, 850 counties reached standard
Taihang Mountain Afforestation Programme	1990–2010	110 counties in 4 provinces	Afforestation of 4 million ha	1990–1999, 3.28 million ha planted
National Programme on Combating Desertification	1991–2000	598 counties in 27 provinces	Control desertification in 7.186 million ha	1991–1999, desertification controlled in 8 million ha
World Bank Loan National Afforestation Project	1990–1997	306 counties in 16 provinces	High-yield fast-growth timber forests	1.39 million ha of new plantation

Source: adapted from Zhang, 2002

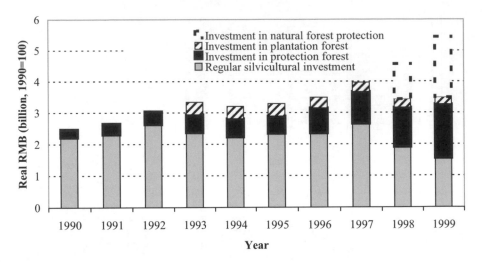

Source: China Statistical Yearbook 2000

Figure 7.3 *Regular state silvicultural investment and project-based investment in China (in billions of real yuan)*

occurred in 1978, when the Green Great Wall was established, and between 1996 and 2001, when the government responded to several large ecological disasters,[10] with the decision to restrict timber harvests from natural forests in some regions and to reforest and protect the upper watersheds.

Figure 7.3 traces the pattern of state investments in selective silvicultural activities since 1990. It shows that investments in large environmental protection and restoration projects have been a rapidly increasing share of the budget. The level of state investment in projects such as these increased more than sixfold during the 1990s.

Effectiveness

How effective were these investments? The level of state investment in forestry has been less of a problem than the effectiveness of the investment. In fact, monies allocated for forest management do not all go to on-the-ground forestry activities. Some are dissipated in the allocation process by forest departments and some are used by SOFEs to support employment, transportation and social services for their employees, as well as for on-the-ground forest management.

Furthermore, once the funding is allocated to actual silvicultural activities, its effectiveness depends upon the incentive structure for SOFE employees. Reforms are gradually replacing the more centralized and planned system of SOFE management with a more decentralized and market-oriented system. Under the old system, silviculture was vertically integrated with logging and wood processing, and silvicultural activities were conducted by individual farms within a state-owned forestry enterprise or by the SOFE itself. During the late 1980s, silviculture, logging and wood processing were divided into separate operations. In addition, the large old forest farms were divided into smaller farms, or into land units now managed by individual households or groups of households.

For silvicultural activities within SOFEs, the first reforms involved wage payments. Wages had been paid for time spent on the task regardless of its level of successful completion. During the mid 1990s, a few SOFEs changed this procedure and began contracting out silvicultural activities. They began tying wages to the amount of work accomplished rather than to the time involved, regardless of output (Zhang, 2001).

A second common arrangement involved a contract under which individuals or small teams performed certain tasks in return for a fixed payment. In all cases, any profits belonged to the contractors responsible for organizing the labour and materials, for site preparation and planting, and for tending the site for the first few years.

A third arrangement was developed to transfer some state-owned equipment to former SOFE employees. In this case, individual workers obtained the right to use equipment in return for contributing a share of the income from its use back to the SOFE.

In a few north-eastern SOFEs, these transfers of land and resource use rights are negotiated or auctioned. The recipients of the new rights are usually limited to SOFE employees. Therefore, the land prices are not entirely determined by the market (Zhang, 2001). In recent years, a growing number of SOFEs have allocated lands as a substitute for salaries and pensions, as well as for severance pay for workers who have been dismissed.

Regardless of the specific contractual reform, the intention has been to link effort with reward. Zhang (2001) assessed the improvements in economic efficiency due to these reforms. He concluded that the reforms resulted in reductions in labour shirking and administrative costs sufficient to improve efficiency in silvicultural activities by 25 per cent in 40 SOFEs in Heilongjiang Province. Others (e.g. Zhang et al, 1994) have reached similar conclusions: silvicultural activities organized in a decentralized system are less costly and achieve better results.

These observations are grounds for optimism. Nevertheless, there remains room for improvement. A recent China Central Television report shed light on the problems of monitoring silvicultural activities (CCTV, 2000). The Shilinji Forest Bureau was devastated by the 1987 forest fire. Over the next ten years it obtained some 27 million yuan of state funds to reforest some 9000ha. The reforestation was completed in 1997. However, an audit found that only 2250ha had been reforested. More generally, the SOFEs have reported an accumulated planted area over the last 20 years that is greater than the area under their management. Yet, SOFEs still contain some wastelands. This means that either the reported number of planted hectares is inaccurate, or many newly planted seedlings have not survived, or both. Future reforms must consider how to improve on these remaining inefficiencies.

Other state forestry activities

Timber production has been the primary objective of the SOFEs. However, the forests also provide a range of non-market environmental services in China.

China is one of the most biodiverse countries in the world with 4400 species of vertebrates[11] (10 per cent of the global total). Its forests contain approximately 8000 species of woody plants, among which are 26 genera of coniferous trees and 260 genera of broadleaves. However, 15–20 per cent of these species are threatened with extinction. Deforestation of the natural forest and the conversion of fragile lands to agriculture are thought to be the most serious causes (Zhang et al, 1994).

An awareness of the need to set up nature reserves for biodiversity conservation has led to the rapid and continuous growth of protected areas in China. China set up its first national forest park, the Zhangjiajie Forest National Park, in 1982. Others were rapidly established for their scenic, wetland or ecological values, or to conserve species. By the end of the 1990s, 1118 nature reserves had been established covering 86.4 million hectares or 8.6 per cent of the national territory (managed by institutions at all levels of government). The eventual target is 10 per cent of all land (BWG, 1997; Harkness 1998; SEPA, 1998; FAO/UN, 2001).

The expansion in nature reserves might be expected to benefit biodiversity, but it has also benefited ecotourism. Ecotourism tends to grow rapidly with increasing personal income and this has been the experience in China. The general tourism data for China show 777,710 person visits in 1982 (China National Bureau of Statistics, 2002), rising to 83 million person visits for forest parks alone in 2001 (SFA, 2002). By 2001, visits to forest parks contributed 37.3 billion yuan of gross revenue (an increase of 25 per cent in only one year) and accounted for 3.5 million jobs.

This is explosive growth, and management problems have accompanied it. The information on the management of protected areas is either limited or anecdotal; but there is evidence that the quality of management is highly variable and generally poor (MacKinnon et al, 1995; SEPA, 1998). One of the major problems is that central government finances released for reserve development are usually 'one-off' investments that do not provide funding for ongoing management operations. The work units responsible for management in each area are encouraged to develop their own economic ventures[12] to meet continuing operating costs, many of which conflict with the declared objectives of protected natural reserves.

Finally, there is watershed protection. In the wake of widespread flooding of the Yangtze and Yellow rivers in 1998, the State Forest Administration (SFA) proposed the Natural Forest Protection Program (NFPP) as a large-scale scheme to protect over 95 million hectares of natural forest by 2010. The State Council has committed 96 billion yuan (US$13 billion) over ten years to finance the programme. Its success can only be determined with time.

Conclusions

China began its transition from a planned economy to a market economy in 1978. In the quarter century since then, private participation has increased in all sectors of the economy. For the forestry sector, the management patterns of the collective forests and the state-owned forests were very similar before 1978. The state provided the majority of investment, and it controlled the procurement and distribution of forest products from both classes of forest. Reform and deviation between the management patterns of the two classifications of forests began when rural households were allowed to contract for the collective forest lands during the late 1970s and early 1980s.

The responses of individual households to reforms in the collective sector were rapid. First the area of forest land and then the volume of standing forest expanded – both substantially. The area and volume of state-owned forests expanded as well, but more slowly, to a lesser extent, and only where there were significant financial inputs from the central government. China's successful experience with household forest management since 1978 should be a lesson for all those countries that are considering some level of transfer of the rights for forest management from the state to local land managers. Farmers can and do manage trees and forests once government removes some of its interference with individual household operations and provides a stable policy environment, and once prices rise to a level that will support forest investment.

Little is known about the contribution of household forestry to rural incomes. However, the evidence from the bamboo sub-sector of forestry suggests that, where there was an effect, forestry income generally was a supplement to income from other sources. Households at all income levels benefited from forest development, but better-off households obtained the greatest benefits (Ruiz Perez et al, 2002).

The experience of the state-owned forests has been different. The central government's role in the management of state-owned forests continued, and its financing of these forests even increased through the mid 1990s. Reform in the state-owned forest sector began more recently with the separation of wood processing and forest management. The wood processing industry is now largely independent. The state-owned share of the processing industry had to become more independent in order to compete with more efficient township and village wood-processing enterprises.

Reform in the state-owned forests themselves is still experimental and restricted to only a few forest bureaus. Auctions of wastelands to private individuals, the increased discretion allowed to managers of SOFEs and the decreasing financial contributions of the central government for capital stock – all occurring since the mid to late 1990s – provide for greater individual opportunity in the state-owned forests. However, the financial difficulties of most state-owned forests, and the willingness of the central government to continue its financial support for silvicultural activities, have delayed any substantial reform.

Prognosis

The larger unresolved questions for the future of forestry in China have to do with the state-owned forests. Their reforms came later and they are still in an experimental stage. They are the larger focus of current policy and they are also closer to the responsibility of this chapter to examine the increasing private role in the activities of state forests.

There are two crucial issues concerning policy for state-owned forests over the next few years and one broader issue for China's full forest sector. The first has to do with making the state-owned forests more market responsive and with allowing them to participate to a greater extent in China's market-driven growth. The central government has promised financial assistance to the SOFEs through to 2010, while they restructure in an effort to become more competitive. It must also allow greater devolution of responsibility to local managers – with fewer constraints on the harvest, sale and shipment of timber.

The second has to do with large-scale forest investments by the central government. The government has shown it can develop and promote large projects such as the Green Great Wall and the more recent Western Regional Development Programme (WRDP) and the Natural Forest Protection Programme (NFPP). Whether these are intended to have direct effects on smallholder forestry operations, or only on state-owned forests, is unclear. Whether programmes such as these are well designed to accomplish their objectives is another matter. In fact, it is not clear that the targets of the WRDP are either the most important causes of erosion or a reasonable means for sustainably improving Western household welfare. The government needs to do a better job of assessing its policy alternatives before embarking on such large projects.

The broader issues have to do with trade and importation of wood and wood fibre. As the NFPP reduces domestic production, China will need to address an imbalance between domestic demand and supply. That imbalance could reach 25 per cent by 2010 (Zhang et al, 1994). This could induce increases in imports, encourage wood substitutes, act as an incentive to improve production efficiency and, ultimately, reduce consumption as prices rise. Whatever its impact, however, an effective NFPP will project substantial future changes in the sources of China's woody raw material.

Joint Management of State Forest Lands: Experiences from India

Sushil Saigal

Introduction

India is a democratic republic with a federal structure, made up of 28 states and 7 union territories spread over 329 million hectares (see Figure 8.1). Climatic conditions vary widely, from permanent snowfields in the Himalayas to tropical coastlands, and from virtual desert in the north-west to fertile, intensively cultivated rice fields in the north-east. India is one of 12 mega-biodiversity countries and contains parts of two global biodiversity 'hotspots'. Forest types range from alpine forests in the Himalayas to rainforests in the Western Ghats. Land degradation and pollution are India's major environmental challenges. More than half of India's area is degraded to some extent (GoI, 1999).

There are more than 1 billion Indians – approximately 80 million of whom are tribal. About 72 per cent of people live in rural areas.

India is the world's twelfth largest economy with an estimated real gross domestic product (GDP) of US$477.3 billion, in 2001. It is the world's fourth largest economy based on purchasing power parity, with an estimated GDP of US$2.93 trillion, in 2001. The annual growth rate of real GDP in 1997–2002 was about 5.4 per cent, one of the highest among major economies in recent years (GoI, 2002a). In per capita terms, however, India ranks a low 128th (2001) (106th on personal purchasing power (PPP) basis) in the world.[1] The agriculture and allied sectors (including forestry) play a key role in the economy, contributing around 24 per cent of GDP (GoI, 2002a) and accounting for 64 per cent of employment (GoI, 1999). Only 27.96 million people work in the formal sector (19.314 million in the public sector and 8.646 million in the private sector) (GoI, 2002a); the vast majority work in the informal sector.

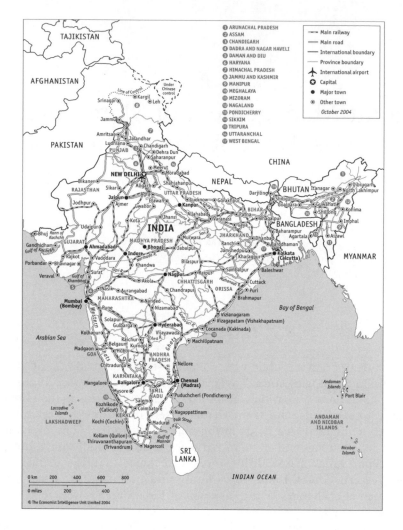

Source: © Reproduced by permission of the Economist Intelligence Unit

Figure 8.1 *India and its states*

Socially, poverty is the biggest challenge. Although the percentage of people living below the poverty line has declined sharply during the last 25 years – 55 per cent in 1973–1974, 36 per cent in 1993–1994 and 26 per cent in 1999–2000 – the absolute number has not. About 260 million people still live in abject poverty. Although only 4.4 and 6.16 per cent of people in Goa and Punjab were living below the poverty line in 1999–2000, in Orissa and Bihar it was 47.15 and 42.60 per cent, respectively. Literacy has increased significantly from 52 per cent in 1991 to 65 per cent in 2001; but around 350 million people are still illiterate (GoI, 2002a).

Major economic reforms in 1991 deregulated the economy to accelerate investment, growth and employment, and hence reduce poverty. India's foreign currency assets grew from less than US$1 billion in 1991 to more than US$45 billion at present. The debt service ratio has declined from a peak of 35.3 per cent of current receipts in 1990–1991 to 16.3 per cent in 2000–2001. The fiscal deficit has reached alarming proportions in many states, however, and it is widely felt that second-generation reforms are urgently needed if India is to accelerate its economic growth.

Overview of the Indian forestry sector

Legal and policy context

Around 23 per cent of the country's area (76.53 million hectares) is classified as forest land.[2] Forestry is the second major land use in the country after agriculture. The land-use pattern is summarized in Table 8.1.

Table 8.1 *Land-use pattern in India*

Land-use category	Area (1991–1992) in million hectares	Percentage
Net area sown (agriculture)	142.50	43.35
Forest (legal)	76.52	23.27
Urban and developmental use	21.88	6.66
Uncultivable wasteland and others	32.83	9.99
Pasture	12.00	3.65
Miscellaneous tree crops	3.00	0.91
Cultivable wasteland	16.00	4.87
Fallow land	24.00	7.30
Total	328.73	100.00

Source: GoI, 1999

The recorded forest area is classified as reserved (55.44 per cent), protected (29.18 per cent) or un-classed forests (16.38 per cent). Reserved forests are fully protected under the Indian Forest Act or State Forest Act, and all activities are prohibited unless permitted. Protected forests have limited protection, and all activities are permitted unless prohibited. Un-classed forests are areas recorded as forest but not reserved or protected, and ownership varies from state to state. There are 87 national parks and 485 wildlife sanctuaries (4.06 million hectares and 11.54 million hectares, respectively). There are 23 tiger reserves spread over 3.3 million hectares. The 11 biosphere reserves (4.76 million hectares) are unique ecosystems chosen for their biodiversity, naturalness and effectiveness as a conservation unit (FSI, 1999).

In addition, 63,618 forest protection committees (FPCs) protect and manage 14.1 million hectares of forest lands under the Joint Forest Management (JFM) programme.

Both central and state governments can legislate on forestry-related matters. State forest departments (FDs) are independent, while national policy is decided centrally by the Ministry of Environment and Forests (MoEF).

The national policy framework is the 1988 National Forest Policy. It stresses managing forests for their environmental and ecological functions and for meeting the subsistence needs of forest fringe people. The national goal is to have at least one third of the country's area under tree cover. However, the policy is only a statement of intention and does not have the force of law.

The legal framework is provided by three main national laws: the 1927 Indian Forest Act, which is the basis for forest administration; the 2002 Wildlife (Conservation) Amendment Act, which governs the 15.6 million hectares of protected area (national parks and sanctuaries); and the 1980 Forest (Conservation) Act (amended in 1988), which mainly controls the diversion of forest land for non-forest purposes. Recently, the 2002 Biological Diversity Act has been passed, which seeks to regulate access to the country's biodiversity. The act has provisions for setting up the National Biodiversity Authority, state biodiversity boards and local biodiversity management committees. The 1986 Environment Protection Act (amended 1991) is the other major legislation that affects forest administration in the country.

There are also numerous state policies, laws, rules and orders covering forest administration, to marketing and transport of forest produce. There are special provisions for the administration of scheduled tribal areas (areas associated with social and ethnic groups recognized in law).

Laws related to the political decentralization process, such as the 1992 Constitution (73rd Amendment) Act and the 1996 *Panchayats* (Extension to the Scheduled Areas) Act, also significantly affect the forestry sector as they give *Panchayati Raj* institutions (PRIs) powers over several forestry-related matters (democratically elected *Panchayati Raj* institutions are the third tier of government, after central and state).

Forestry in the national economy

Contribution to GDP and employment

While forests occupy 23 per cent of the land area, forestry and logging contributed a mere 1.07 per cent of GDP in 2000–2001 (constant prices), down from 4.42 per cent in 1970–1971 (CMIE, 2003). However, forest-based industries are included in the 'manufacturing sector'; more importantly, the GDP figures are based on recorded removals from forests – a fraction of actual removals – and do not take into account the environmental services provided by forests. Furthermore, the contribution of many farm foresters is perhaps recorded under agriculture rather than forestry. This has resulted in a substantial underestimation of the contribution of forestry to the national economy.

Forestry creates jobs. Around 70 per cent of the plantation budget is direct wages. About 250 million person days of employment are generated each year by forestry development schemes in state forests and other government lands. Protection, maintenance and harvesting generate about 100 million person days annually and agroforestry and farm forestry another 75 million (GoI, 1999).

Forestry and livelihoods[3]

That forestry helps to provide livelihoods – especially to marginalized and vulnerable people – is not widely recognized. India has perhaps the largest number of poor (approximately 260 million) and indigenous peoples (approximately 80 million) in the world. Many live in or near forests and maps of forests, poverty and indigenous peoples overlap considerably (Poffenberger and McGean, 1996). About 147 million people live near forests (FSI, 1999).

Many of these people depend upon forests for fuelwood, fodder, small timber for agricultural implements and house construction, and even food and medicines – otherwise known as non-timber forest products (NTFPs). Selling fuelwood and NTFPs also generates vital income. Dependence is greatest among the very poor, such as landless workers. In 1981, one in nine rural households did not own any land (Hague, 1987, in Mishra, 1997), a situation that has probably deteriorated even further now. Even those with small landholdings often need to earn a wage as well, and jobs are scarce, so many depend upon forests and village common lands for survival.

According to India's submission to the United Nations Conference on Environment and Development (UNCED) in Rio in 1992, 70 per cent of rural and 50 per cent of urban people use fuelwood for cooking (Dwivedi, 1993; World Bank, 1993). More than 100 million animals graze on forest lands (Dwivedi, 1993; World Conservation Union, 1991, in WRI et al, 1994).

A survey of 170 households in Bihar showed that for 20 per cent, selling fuelwood was a major source of income (World Bank, 1993). Millions collect *tendu* (*Diospyros melanoxylon*) leaves to make cigarettes (*bidis*). In Uttar Pradesh about 200,000 people depend upon harvesting and processing *bhabbar* grass (*Eulaliopsis binata*) (Poffenberger and Sarin, 1995). In West Bengal, 72 per cent of households stitch leaf plates from *sal* (*Shorea robusta*) leaves, generating nearly half of their households' income (Dutta and Adhikari, 1991). For many women in Orissa, collecting NTFPs is their primary occupation. Households on less than 3000 rupees derived 50 per cent of their earnings from NTFPs (Malik, 1994). Some 600 million tonnes of forest products valued at 300 billion rupees are collected annually from India's forests (GoI, 1999).[4]

Forest-based industries

Some government enterprises were started after independence; but most forest product processing is carried out by the private sector, including more than 90 per cent of India's wood-based products (GoI, 1999).

Most processing is in small units. Some 98 per cent of the 23,000 saw mills are small units with an annual log intake of 3000 cubic metres (GoI, 1999).

The actual number of forest-based enterprises is unknown – in Yamuna Nagar and Rajkot districts, as many as 27 per cent and 98 per cent, respectively, of enterprises were unregistered (Saigal et al, 2002).

Although assessments vary widely, the government estimates that total consumption of wood by the processing industries is 24 to 30 million cubic metres per year (GoI, 1999). The pulp and paper industry produces paper and newsprint worth 90 billion rupees every year and pays 16 billion rupees in excise duty and taxes (Singhania, 1997).

During 1997–1998, India exported forest products (selected groups) worth 36.3 billion rupees and imported forest products worth 60.3 billion rupees (ICFRE, 2000).

Farm and agroforestry

Farmers now supply about 50 per cent of wood (GoI, 1999), as supplies from state forests have declined because of conservation and a green felling ban in several states. Farmers mainly grow block or field bund plantations of commercially valuable fast-growing species, such as eucalyptus, poplar, acacia, casuarina and *Leucaena*.

Farmers in Prakasam and Uddham Singh Nagar alone produce nearly 1 million metric tonnes of wood valued at 1.2 billion rupees annually (Saigal et al, 2002).

Investment in the forestry sector

Investment in the forestry sector is low. While forest produce worth 300 billion rupees is extracted annually from the government forests alone, total investment (government and private) is less than 15 per cent of this amount, or 41.7 billion rupees (GoI, 1999). Expenditure in Five-Year Plans has usually been less than 1 per cent of the total outlay, and the proportion of forest budget to GDP has declined over the past decade (Kadekodi, 2001).

Current status of the resource and its management

Although 23 per cent of India (76.53 million hectares) is officially classified as forest land, not all of it has tree cover. At least 17 per cent has almost no tree cover (crown density below 10 per cent),[5] another 33 per cent is degraded open forest (crown density 10–40 per cent), and less than half has reasonably dense forest (crown density above 40 per cent) (FSI, 1999). The status of forests is summarized in Table 8.2.

Forest cover is concentrated in the north-east, the Himalayas, the central tribal belt, the western and eastern Ghats, and patches of coastal mangroves (see Table 8.3).

Most of the forest lands are owned and controlled by the government, and are divided into reserved, protected and un-classed forests (see Table 8.4).[6] While most reserved and protected forests are controlled by state forest departments, a significant proportion of un-classed forests, especially in the north-east, are managed by community groups. During recent years, FDs have started involving local communities in the protection and/or management of forests under the JFM programme.

Table 8.2 *Status of India's forests*

Category	Area (million hectares)	Percentage of total geographical area	Percentage of forest area
Total geographical area	328.73	100.00	–
Legally classified forest	76.53	23.28	100.00
Actual forest cover [a]	63.73	19.39	83.27
Open forest [b]	25.51	7.76	33.33
Dense forest [c]	37.73	11.48	49.30
Mangroves	0.49	0.15	0.64
Plantations	34	10.34	–

Notes:
a Forest with at least 10 per cent crown cover
b Forest with crown cover between 10 and 40 per cent
c Forest with over 40 per cent crown cover

Source: FSI, 1999; GoI, 1999

Demand-and-supply scenario[7]

There is a huge gap between the demand and sustainable supply of forest products. In terms of volume, fuelwood is the most important product of India's forests, making up 80 per cent of total national demand for wood. Between 210 and 233 million tonnes were needed in 2001; but annual allowable (sustainable) cut from government forests is only 44 million tonnes and from private forests 35 million tonnes – a gap of some 131 million to 154 million tonnes. This gap is being filled through unsustainable removals and unrecorded farm forestry and common land sources.

Table 8.3 *Forest cover in selected states and union territories of India*

States/union territories	Forest cover (approximate percentage)
Arunachal Pradesh, Mizoram, Nagaland and Andaman and Nicobar Islands	>80
Madhya Pradesh*, Orissa, Goa and Assam	30
Kerala and Himachal Pradesh	25
Karnataka, Andhra Pradesh, Bihar*, Maharashtra and Tamil Nadu	15
Uttar Pradesh* and West Bengal	10
Rajasthan, Punjab and Haryana	<5

Note:
*These states have now been divided; the figures are for undivided states.

Source: FSI, 1999

Table 8.4 *Status of India's forests by ownership and management categories*

Type of forest	Area (million hectares)
Reserved forest	41.65
Protected forest	22.33
Un-classed forest	12.53
Protected areas	15.60
Plantations of state forest corporations	1.24
Joint Forest Management (JFM) forests	14.10

Source: FSI, 1999; GoI, 1990, 2002b

Industrial wood – that is, everything except fuelwood – is a small part of total demand but is also in short supply. Overall demand is 50 million to 65 million cubic metres for 1996, while the annual allowable cut is only 26 million to 27 million cubic metres – a shortfall or unsustainable removal of 23 million to 38 million cubic metres.

Key players

The government is the dominant player in the forestry sector in India, owning and controlling about 97 per cent of all forests, while just 1.13 per cent is owned by communities and corporate bodies and 1.55 per cent comprises private forests (ICFRE, 2000).[8] Government also closely regulates forestry on non-forest land and plays a significant role in afforestation.

Although 90 per cent of wood-based products are manufactured in the private sector, for legal and political reasons it does not play an important direct role in plantations. However, recent information from 12 companies indicates that they are supplying more than 53 million seedlings annually to farmers, covering an estimated 26,000 hectares, a quarter of which is of clonal origin (Saigal et al, 2002).

Farmers are now important suppliers of wood, and communities have begun protecting and managing forests through the JFM programme: 63,618 communities protect and regenerate more than 14 million hectares (or over 18 per cent) of government forest lands under JFM.

The non-governmental organization (NGO) community is diverse in both size and area of operation. Large NGOs such as the Centre for Science and Environment, World Wide Fund for Nature-India (WWF-India) and the Society for Promotion of Wastelands Development have successfully lobbied against the leasing of state forest lands to industry for plantations.

During the past few years, the judiciary has started playing an important and proactive role in matters related to the environment. The judgements of the courts in several public interest litigation cases have virtually set the policy.

External funding agencies also influence forestry policy, especially where they provide loans or grants.

Commercial forestry and plantations: Historical perspective and current status

Historical perspective

Although earliest available records of forestry date back to the Maurya Empire (circa 300 BC), systematic forest management only began in the colonial period (Lal, 1992).

During the early nineteenth century, forests began to supply timber for ship-building, local construction and export and, later, an extensive railway network. The rate of deforestation must have been alarming even then, as the India Navy Board stressed the need for conservation policies as early as 1830 to save the forests from devastation (Hobley, 1996).

The government soon began creating plantations, the first being a native teak plantation at Nilambur, Kerala, in 1840 (GoI, 1999).

In 1864, the Indian Imperial Forest Service was set up to manage the country's forests. Regular planting, mainly of teak, started in 1865 in the teak-growing central and southern provinces. Eucalyptus was introduced to the Nilgiri Hills of the present Tamil Nadu State in 1858. Small-scale planting began of commercially valuable species such as rosewood, mahogany, *toon* (*Cedrela toona*) and *sal*, partly for research purposes, including provenance trials and growth studies. Plantation activity received a boost in 1911 when the *taungya* system was introduced, whereby crops are grown between the rows of trees during the initial years of the plantation before the tree canopy closes (FSI, 1999; GoI, 1999), and after World War I. These plantations were often established after clearing mixed natural forests (Hobley, 1996). Most planting took place on lands over which the government had declared its proprietary rights through the forest acts issued in 1865, 1878 and 1927. The rights of the local communities over these forests were significantly curtailed (Singh, 1986; Vira, 1995).

Industrial and revenue considerations dominated Indian forestry after independence. During the first 25 years of development planning (1951–1974), commercial forestry accounted for over 65 per cent of physical area coverage and over 80 per cent of the total financial outlay for afforestation (Vira, 1995). As early as 1961 it was estimated that the shortfall in fuelwood supply by 1975 would be around 100 million tonnes, and for industrial wood 4 million tonnes. Yet, three-quarters of the plan's afforestation was for commercial plantations (Vira, 1995).

Most industrial plantations until 1979 were established on state forest lands after the clear-felling of 'low-value' native forests. Most plantations were teak, *sal*, *deodar* (*Cedrus deodara*), chir pine (*Pinus roxburghii*), eucalyptus and acacias.

A report of the National Commission on Agriculture (NCA) in 1976 led to increased investment in plantation establishment. In order to enhance productivity and employment, the report recommended the large-scale replacement of mixed natural forests of low commercial value with fast-growing commercially important plantation species. It also recommended that 48 million hectares of state forest land should be committed to production forestry (GoI, 1976), and that forest development corporations (FDCs) be created to manage forests on business principles and to attract finance (see Box 8.1).

Box 8.1 India's forest development corporations

Most forest development corporations (FDCs), were created during the 1970s and 1980s to undertake production forestry. The National Commission on Agriculture (NCA) had recommended that 48 million hectares of forest land should be dedicated to production forestry, and its interim report (August 1972) set the following objectives:

- Raise both volume and value per hectare.
- Create both skilled and unskilled jobs.
- Support substantially the economy of under-developed areas and tribal people.
- Expand or establish many forest-based industries.
- Begin exporting wood and wood products.
- Have a sustainable impact on employment in the secondary and tertiary sectors.

By 1990, there were 26 FDCs in 20 states, established as follows:

- 1962: Orissa Forest Corporation;
- 1969: Forest Development Board of Maharashtra (converted into Forest Development Corporation of Maharashtra in 1974);
- 1971: Mysore Forest Plantation Corporation (later renamed Kartanataka Forest Plantation Corporation, then, in 1987, Kartanataka Forest Development Corporation);
- 1973: Kartanataka State Forest Industries Corporation;
- 1974: Uttar Pradesh Forest Corporation; Himachal Pradesh State Forest Corporation; Tamil Nadu Forest Plantation Corporation; and West Bengal Forest Development Corporation;
- 1975: Kerala Forest Development Corporation; Madhya Pradesh State Forest Development Corporation; Forest Development Corporation of Meghalaya; Bihar State Forest Development Corporation; Andhra Pradesh Forest Development Corporation; and Tamil Nadu Tea Plantation Corporation;
- 1976: Gujarat State Forest Development Corporation; and Tripura Forest Development and Plantation Corporation;
- 1977: Andaman and Nicobar Islands Forest and Plantation Development Corporation; and Arunachal Pradesh Forest Corporation;

- 1978: Karnataka Cashew Development Corporation;
- 1979: Jammu and Kashmir Forest Corporation;
- 1980: Simlipahar Forest Development Corporation (Orissa);
- 1983: Punjab Forest Development Corporation;
- 1984: Arasu Rubber Corporation, Tamil Nadu;
- 1985: Rajasthan State Forest Development Corporation;
- 1990: Haryana State Forest Development Corporation (GoI, 1990).

Some FDCs are only involved in harvesting, processing or marketing produce on behalf of the forestry department (FD) or farmers.

By 1990 the total area of state-owned plantations controlled by FDCs was 1,236,487ha (GoI, 1990). Eleven FDCs had created 966,538ha of teak, eucalyptus, bamboo, pine and casuarina plantations. The FDCs of Karnataka, Andhra Pradesh, Tamil Nadu and Kerala had acquired 251,174ha of existing forest plantations from the FDs. Five other FDCs had established 18,774ha of cash crop plantations (red oil palm, rubber, cashew, tea, cardamom, coffee and lavender).

However, the FDCs review in 1990 noted that FDC plantations were not very successful. Plantations were, on average, 55 to 60 per cent successful (probably low seedling survival rate). Their plantations had low yields, poor growth and low survival rates, and the quality of the products was very poor (GoI, 1990).

Most FDC plantations were incurring losses and could not attract institutional finance.

The main reasons for the poor performance of FDCs are listed below (adapted from GoI, 1990):

- *Excess state control*: FDCs have hardly any functional autonomy to manage their affairs – technical or commercial.
- *Lack of corporate culture*: most FDCs work as a government department, with top positions appointed by the FD alone. The managing director (MD) is normally an Indian Forest Service bureaucrat.
- *Lack of long-term vision*: the average tenure of Managing Directors was only 20 months; hence, long-term planning and vision were poor.
- *Diversification into unrelated activities*: Bihar FDC built 425 primary schools and was running an *Ayurvedic* dispensary. Karnataka FDC built tourist cottages in a wildlife sanctuary. Many FDCs in the south of India were engaged in the rehabilitation of Sri Lankan refugees.
- *Excess staff*: most FDCs are overstaffed.

Source: Saigal (2000)

The NCA viewed local communities' dependence on forests as a major cause of forest destruction and a major obstacle for production forestry. It suggested that local communities' needs be met by a social forestry programme on non-forest lands, such as village commons, government wastelands and farmlands

(GoI, 1976). This led to the birth of a social forestry programme and many plantations were established near villages.

During the mid 1970s and 1980s, many FDCs were established, the social forestry programme was launched and a massive afforestation effort began along the lines of the NCA report (see Figure 8.2). The social forestry plantations were primarily of fast-growing species for fuelwood, poles and inferior timber, and on government lands these were established and managed by the FDs with little involvement of local people. Plantations on village common lands were often established and managed for a few years by the FDs and then handed over to *Panchayati Raj* institutions (PRIs). The plantations on private lands were established by farmers mainly using either free or highly subsidized seedlings supplied by the FDs. Funds for the social forestry plantations came primarily from donors.[9]

The annual planting rate increased to about 1 million hectares per year during 1980–1985. The National Wastelands Development Board (NWDB) was established in 1985 and given a target of afforesting 5 million hectares of wastelands per annum. Although NWDB did not quite achieve its ambitious target, it helped the social forestry programme by providing additional funds for plantations. It also got NGOs involved in the government's afforestation effort in a significant way for the first time.

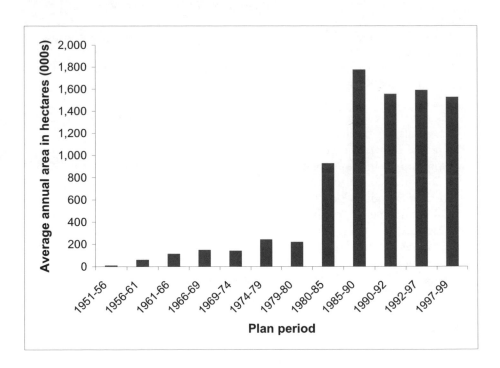

Source: FSI, 1999

Figure 8.2 *Annual rate of plantation development in India*

A record 8.86 million hectares were established during 1985–1990 (FSI, 1999). Investment had increased after the creation of the NWDB for tree plantations under the poverty alleviation schemes (25 per cent of funds under the National Rural Employment Programme and the Rural Landless Employment Guarantee Programme were earmarked for planting purposes). In the Seventh Plan, 25.87 billion rupees were invested – more than double the total cumulative investment of 11.67 rupees up until then (GoI, 1999).

Annual plantation rate has slightly declined since 1991 and is about 1.5 million hectares. Between 1951 and 1999, 31.20 million hectares have been planted (FSI, 1999). It is important to note that a significant proportion of the estimated area of plantations has been calculated on the basis of seedlings distributed, rather than actual area planted.

Cumulative plantation area by state at the end of 1999 is presented in Table 8.5.

Current status

It is very difficult to get an accurate estimate of the total plantation estate and its division between forest and non-forest lands. It is not known how many of the plantations included in the cumulative estimate above were unsuccessful, how many have been harvested and how many replanted. Furthermore, Table 8.5 does not include plantations established by farmers without government seedlings. The National Forestry Action Programme report does provide an estimate of the age distribution of plantations at the end of 1996, based on plan targets and achievements (see Table 8.6).

According to these estimates, the plantation estate at the end of 1996 was between 19.5 million and 28 million hectares. Of this, 6 million to 8 million hectares are estimated to be outside the state forest lands (GoI, 1999). According to the Forest Survey of India, 15.3 million hectares of plantations were established by FDs until 1997 (see Table 8.7). In its Global Forest Resource Assessment (FAO/UN, 2001) the United Nations Food and Agriculture Organization (FAO) estimates that India has 18 per cent of the world's plantations, or about 34 million hectares.

As indicated in Table 8.7, the main species in plantations established by FDs are eucalyptus (8.87 per cent), teak (8.67 per cent) and *Acacia nilotica* (5.23 per cent). Another notable feature is that the 'others' category constitutes 58.28 per cent.

Evolving demands and changing roles[10]

Colonial period

The colonial government managed the forests for timber production. Although the rights of local communities were significantly curtailed, forest policy was not entirely unsympathetic:

Table 8.5 *Cumulative area of India's forest plantations by all agencies in all states and union territories, 1951–1999 (thousand hectares)*

State/union territory	Total cumulative plantation area	Area of block plantations	Area converted from seedlings
Andhra Pradesh	2496.56	1260.30	1236.26
Arunachal Pradesh	160.95	155.73	5.22
Assam	451.78	433.58	18.20
Bihar	1326.23	942.12	384.11
Goa	65.60	46.04	19.56
Gujarat	2981.08	1293.95	1687.13
Haryana	742.74	597.02	145.72
Himachal Pradesh	719.44	665.84	53.60
Jammu and Kashmir	382.43	323.04	59.39
Karnataka	2163.22	1573.19	590.03
Kerala	688.12	483.63	204.49
Madhya Pradesh	3364.13	2848.52	515.61
Maharashtra	2965.07	2130.39	834.68
Manipur	154.76	139.69	15.07
Meghalaya	164.48	130.67	33.81
Mizoram	308.55	255.73	52.82
Nagaland	174.20	116.43	57.77
Orissa	1827.41	1458.49	368.92
Punjab	512.38	417.60	94.78
Rajasthan	1410.10	1150.79	259.31
Sikkim	119.23	107.53	11.70
Tamilnadu	2268.18	1616.18	652.00
Tripura	246.64	215.61	31.03
Uttar Pradesh	4185.77	1844.36	2341.41
West Bengal	1157.73	610.93	546.80
Andaman and Nicobar Islands	88.14	83.13	5.01
Chandigarh	10.07	9.85	0.22
Dadra and Nagar Haveli	18.36	11.01	7.35
Daman and Diu	1.39	0.85	0.54
Delhi	44.05	20.18	23.87
Lakshadweep	2.50	0.57	1.93
Pondicherry	7.88	1.91	5.97
Total	31,209.17	20,944.86	*10,264.31

Note:

* The area has been estimated by the Forest Survey of India using 1990–1999 figures of the National Afforestation and Ecodevelopment Board, where break-up of block plantations and seedlings distributed are available.

Source: FSI, 1999

Table 8.6 *Age-class distribution of India's forest plantations*

Age class (years)	Estimate (million hectares)
0–5	5.0–8.0
5–10	8.0–10.5
10–15	4.0–6.0
Above 15	2.5–3.5

Source: GoI, 1999

Table 8.7 *Plantations established by India's forest departments until 1997, by species*

Species	Area in thousand hectares	Percentage
Eucalyptus spp.	1,360.91	8.87
Tectona grandis	1,330.09	8.67
Acacia nilotica	801.61	5.23
Acacia auriculiformis	564.67	3.68
Bamboo	408.09	2.66
Pinus roxburghii	318.54	2.08
Dalbergia sissoo	266.58	1.74
Acacia catechu	259.54	1.69
Shorea robusta	250.28	1.63
Gmelina arborea	148.01	0.97
Anacardium occidentale	141.54	0.92
Casurina equisetifolia	133.99	0.87
Pinus kesiya	127.12	0.83
Cedrus deodara	124.93	0.81
Populus spp.	47.48	0.31
Bombax ceiba	37.97	0.25
Acacia mearnsii	37.56	0.24
Picea smithiana, Abies pindrow	16.74	0.11
Hevea brasiliensis	12.30	0.08
Santalam album	10.58	0.07
Others	8,938.10	58.28
Total	15,336.63	100.00

Source: FSI, 1999

> *Every reasonable facility should be afforded to the people concerned for the full and easy satisfaction of these (subsistence) needs, if not free, then at low and not at competitive rates. It should be distinctly understood that considerations of forest income are to be subordinated to that satisfaction (GoI, 1894).*

The conversion of forest to agriculture (and, hence, revenue-paying land) was also encouraged:

> *...wherever an effective demand for culturable land exists and can only be supplied from forest area, the land should ordinarily be relinquished without hesitation (GoI, 1894).*

Post-independence period

A new forest policy in 1952 was much less sympathetic to communities' needs:

> *The accident of a village being situated close to the forest doesn't prejudice the right of the country, as a whole, to receive the benefits of a national asset (GoI, 1952).*

During the 1950s, India was focusing on self-sufficiency in food production and the building of indigenous industrial infrastructure. Forests were raw material and sources of revenue (Vira, 1995). Substantial capital was invested in forest-based industries (Gadgil et al, 1983, in Hobley, 1996). Bamboo consumption went up from 58,000 tonnes per year in 1945 to about 5 million tonnes in 1987 (Hobley, 1996). In order to encourage industrial development, many forest-based industries were subsidized or given cheap long-term leases.

National Commission on Agriculture

The National Commission on Agriculture (NCA) report greatly influenced forestry during the 1970s and 1980s. Increasing concern about low productivity led the NCA to suggest that large-scale industrial plantations should be established on state forest lands. The report noted that while forests occupied 23 per cent of India's landmass, their contribution to the gross national product (GNP) was less than 1 per cent (GoI, 1976), ignoring both the non-monetized forest-based economy of rural and tribal communities and the economic value of the protective functions of the forests.

When the newly created forest development corporations (FDCs) began establishing commercial plantations after clearing mixed natural forests, there was stiff local resistance. Rural people, especially tribal communities who depended upon mixed natural forests for subsistence and had customary access to these forests, protested against the destruction. In the Bastar District of Madhya Pradesh, a World Bank-supported project to replace 20,000 hectares of native mixed *sal* forest with tropical pines in 1975 had to be dropped after protests by local tribal communities (Dogra 1985; Anderson and Huber, 1988; Pathak, 1994). In Bihar, there were protests against replacement of natural forests with teak plantations (CSE, 1982).

The NCA report also paved the way for social forestry plantations on non-forest lands. The initial objective was to reduce the dependence of local communities on state forest lands so that these lands could be committed to production forestry. The NCA's other concerns were to develop wastelands and increase agricultural productivity. It was hoped that this increased supply of fuelwood would meet local needs and even generate a surplus. This, in turn, would reduce the use of cow dung as fuel so that it could then be used as manure.[11]

Social forestry got further support from the environmental lobby within the central government, who hoped it might solve land degradation. The energy crisis in the mid 1970s encouraged international assistance, as social forestry projects were meeting fuel needs (World Bank, 1983, in Pathak, 1994). This support was based on the assumption that the 'real energy crisis' for more than one third of the world's population was the daily scramble to cook dinner (Eckholm, 1975, in Rao et al, 1992).

It was during the Sixth Five-Year Plan (1980–1981 to 1984–1985) that social forestry acquired the 'people-oriented' forestry connotation (Pathak, 1994) and became a significant part of the government's rural development and employment generation efforts. A number of new programmes were started with social forestry as an important component (Vira, 1995).

At the same time, there was growing concern about the continuing degradation of forests, which led to the 1972 Wildlife (Protection) Act and the 1980 Forest (Conservation) Act. While many protected areas were created under the Wildlife (Protection) Act, the Forest (Conservation) Act ensured that state governments needed the approval of central government before converting forest land to any non-forest use.

New forest policy

By the mid 1980s, it was obvious that the NCA strategy was not working: there was continuing forest degradation and increasing conflicts between the local communities and FDs. The link between environment degradation and poverty was also better appreciated by policy-makers.

The following quotes from the policy document clearly show that the 1988 policy is radically different from the earlier approach, focusing on conservation and communities' needs:

> *The principle aim of forest policy must be to ensure environmental stability and maintenance of ecological balance, including atmospheric equilibrium, which are vital for sustenance of all life forms, human, animal and plant. The derivation of direct economic benefit must be subordinated to this principal aim (GoI, 1988).*
>
> *The life of tribals and other poor living within and near forests revolves around forests. The rights and concessions enjoyed by them should be fully protected. Their domestic requirements of fuelwood, fodder, minor forest produce*

and construction timber should be the first charge on forest produce (GoI, 1988).

The practice of supply of forest produce to industry at concessional prices should cease. Industry should be encouraged to use alternative raw materials. Import of wood and wood products should be liberalized (GoI, 1988).

Natural forests serve as a gene pool resource and help to maintain ecological balance. Such forests will not, therefore, be made available to industries for undertaking plantation and for any other activities (GoI, 1988).

No such [plantation] programme, however, should entail clear-felling of adequately stocked natural forests. Nor should exotic species be introduced, through public or private sources, unless long-term scientific trials undertaken by specialists in ecology, forestry and agriculture have established that they are suitable and have no adverse impact on native vegetation and the environment (GoI, 1988).

A 1988 amendment to the Forest (Conservation) Act prohibited the clearing of natural forests for plantations, and central government would have to approve any leasing arrangements to the private sector.

Joint Forest Management

In 1990, based on encouraging results from pioneering experiments in community-based forest management, the government started the Joint Forest Management (JFM) programme. A circular (No 6.21/89-FP, dated 1 June 1990) directed all states to involve local communities and voluntary agencies in protecting, afforesting and developing degraded forest lands.

Under JFM, the forest department and the community agree to jointly protect and manage forest land adjoining villages and to share responsibilities and benefits. The community is represented through a body specifically formed for the purpose.[12]

The villagers have access to NTFPs and a share in timber revenue in return for protecting the forests from fire, grazing and illicit harvesting. The details vary from state to state; but in all states, ownership of the land remains with the government.

JFM spread countrywide during the 1990s and community groups emerged as important managers of state forests and plantations. By March 2002, there were 63,618 groups protecting and managing over 14 million hectares (18 per cent) of state forest lands. The breakdown by states is given in Table 8.8.

Panchayati Raj institutions

In 1992, the constitution of India was amended and *Panchayati Raj* institutions (PRIs) were given a mandate to prepare and implement plans for economic development and social justice on 29 subjects, including social forestry, farm forestry, minor forest produce, fuel and fodder.

In 1996, this amendment was made applicable to Schedule V areas of the country.[13] The 1996 *Panchayats* (Extension to the Scheduled Areas) Act, among others, granted ownership rights over minor forest products to PRIs.

Table 8.8 *Progress of Joint Forest Management in India from 1 March 2002*

State number	States	Area under JFM (square kilometres)	Number of forest protection committees
1	Andhra Pradesh	17,675.70	6,816
2	Arunachal Pradesh	58.10	13
3	Assam	69.70	245
4	Bihar	741.40	296
5	Chhattisgarh	28,382.55	6,412
6	Goa	130.00	26
7	Gujarat	1,380.15	1,237
8	Haryana	658.52	471
9	Himachal Pradesh	1,112.47	914
10	Jammu & Kashmir	795.46	1,895
11	Jharkhand	4,304.63	1,379
12	Karnataka	1,850.00	2,620
13	Kerala	49.95	32
14	Madhya Pradesh	43,000.00	10,443
15	Maharashtra	6,866.88	2,153
16	Manipur	5,072.92	82
17	Mizoram	127.40	129
18	Nagaland	1,500.00	55
19	Orissa	7,834.67	12,317
20	Punjab	735.60	184
21	Rajasthan	3,093.36	3,042
22	Sikkim	6.00	158
23	Tamil Nadu	3,733.89	999
24	Tripura	319.89	180
25	Uttar Pradesh	507.03	540
26	Uttaranchal	6,066.08	7,435
27	West Bengal	4,880.95	3,545
	Total	140,953.30	63,618

Source: GoI, 2002b

Changing roles of the private sector: Achievements and challenges

Role of forest fringe communities: Joint Forest Management

Since the 1988 forest policy was issued, protecting the forest's environmental services and meeting the subsistence needs of the forest fringe communities have been the main stated objectives of forest management.

The JFM programme began after policy-makers realized that it is impossible to manage forests successfully without the active and willing participation of forest fringe communities. Furthermore, these communities should benefit directly and have sufficient authority to be effective. From the planners' perspective, JFM's focus on natural regeneration is cheap compared with new plantations.

JFM's main positive impacts include the following (GoI, 2002b).

Improvement in the condition of forests: Overall forest cover has increased by 3896 square kilometres and dense forest cover by 10,098 square kilometres. In JFM areas, illegal felling has declined sharply, even during droughts.

Increase in income. In six states alone, JFM generated 21.58 million person days of employment in 2001–2002. The Andhra Pradesh Forestry Project (1994–2000) created over 40 million days of employment. In Madhya Pradesh, JFM irrigation facilities have increased the yield by two to five times. In Gujarat, better availability of grass and tree fodder has increased milk production – in Nisana village (Vyara Division) from 40,000 to 200,000 litres per year.

In four states, forest protection committees (FPCs) received 62.59 million rupees through benefit-sharing under JFM during 2001–2002. Income from NTFPs is generally more than the community's share in timber revenue. Some women in West Bengal earn 4500 to 6000 rupees annually selling *sal* leaf plates.

Change in attitude and relationship. One of the most significant impacts of JFM has been the change in attitudes of local communities and forest officials towards each other and forests. One FPC postponed a village wedding to fight a forest fire. Some traditional forest protection practices have been revived – for example, *kesar chhanta* (sacred groves) in Rajasthan.

However, in most states JFM still depends upon donor funds, and programmes can be changed or withdrawn by FDs at any time: there is no security of tenure for communities. The communities' share of revenue from forests is still low in many states and the collection and sale of several commercially valuable NTFPs are restricted. Another key challenge is to get the right links between PRIs and FPCs, where there is currently considerable confusion.

Marketing the produce from JFM forests is critical. In many cases the market is smaller than the harvest. During 1996–1997, only one fifth of West Bengal's potential JFM harvest could be sold (Guhathakurta and Roy, 2000, in Saigal et al, 2002) due to gluts and a range of competition. The price of particular *sal* poles fell from 166 rupees in 1992 to 140 rupees in 1998; during the same time period, similar eucalyptus poles fell from 110 rupees to 85 rupees (TERI, 1999, in Saigal et al, 2002).

In Haryana, the main NFTP is *bhabbar* grass (*Eulaliopsis binata*); but Ballarpur Industries Limited – the largest buyer – has changed its technology and does not need *bhabbar* in large quantities any longer. The withdrawal of excise concessions on *bhabbar* has made it even less unattractive (Saigal et al, 2002).

Corporate-sector involvement on state forest lands

The corporate sector's involvement on state forest lands is a subject of heated debate. It cannot establish large-scale plantations on non-forest lands due to land ceiling restrictions and the out-grower experiments were not very positive (see Box 8.2).

Many wood-based industries, especially paper and pulp companies, have been lobbying to lease degraded forest lands to establish plantations. This is strongly opposed by some NGOs and environmental action groups because it is against the National Forest Policy and Forest (Conservation) Act and because they feel that companies are likely to overexploit the resource to the cost of local communities. The poor track record of several companies and forest contractors is cited by many as a good enough reason to resist.

Indian companies argue that they have to increase the size of their units in order to remain competitive; but they are unwilling to invest in new plants (e.g. paper mills) until about half of their future raw material supply is ensured. They consider farm forestry supplies unreliable, and want at least half of their supplies from captive plantations or government supplies.

In 1998 the Planning Commission of India formed a working group to examine this issue and strongly recommended against leasing state forest lands to industry, citing a number of reasons:

- It is against the interests of farm foresters because companies will reduce or stop purchasing farm-grown wood.
- It will not lead to additional supply. Cheaper production on farm lands will be replaced by socially more costly production on forest lands.
- Degraded forests satisfy the fuelwood and fodder needs of many people. These lands need protection and recuperation, not industrial development.
- The current market for pulpwood is distorted because of continuing subsidies to industry, especially for bamboo. Industry's attempt to get free access to forest lands without paying the full market price will further distort the market. Heavy subsidies will increase corruption. Subsidizing the rich at the expense of tribal and forest-dependent communities will attract public criticism, and may even give rise to militant movements.
- Paper and other large industries consume just a fraction of forest products. Some 90 per cent of forest raw material is processed by 25,000 saw mills and a larger number of cottage units, who would also lay claims to forests if large industries were able to snatch concessions from government.
- There will be similar claims from the coffee, cashew and palm plantation industries. Like the paper industry, they will establish short-term and quick-growing species in place of the multi-layer mixed forests resulting from natural regeneration, ignoring the ecological implications.

Box 8.2 Constraints in establishing plantations on non-forest lands in India

The private sector is unable to establish large-scale tree plantations on non-forest lands because of statutory land ceilings on agricultural land.[1] The ceiling limits for companies are the same as those for individuals, and the permissible landholdings are simply too small.

Many companies have tried to promote forest plantations on private farms in a number of ways (often with buyback guarantees for the crop):

- supplying free or subsidized seedlings;
- helping farmers to get bank loans;
- establishing and maintaining a plantation while paying the farmer a fee or share of profits;
- investing in the research and development and sale of improved clonal planting stock.

The partnerships were not successful. Today, most companies simply produce and supply seedlings to farmers without a partnership because:

- Farmers did not care for the subsidized seedlings.
- The loan procedures were cumbersome.
- Farmers did not understand the agreements.
- Farmers spent the loans elsewhere.
- Companies did not buy all raw material despite buy-back guarantees.
- Legal frameworks governing leasing and share cropping were unclear, especially in tribal areas.
- There was a fear of litigation.

If some of these problems could be solved, many more company–farmer partnerships could emerge in the country. Loan sanction procedures need to be simplified, and greater clarity is required in the legal framework that governs plantations on the lands of tribal farmers in Schedule V areas, so that companies do not become embroiled in unnecessary litigation.

To increase the corporate sector's direct involvement in managing forest plantations on non-forest lands, the land ceiling restrictions must be reviewed. Other plantation crops such as tea, coffee and bananas receive special exemptions. This could start with an exemption for plantations on private wastelands.[2]

Notes:
1 Land ceilings is one element of land reforms that seem to alleviate poverty and lead to growth with distributional equity. The actual limit varies between states; but total holdings should not exceed 54 acres (IASSI, 1991).
2 There are some schemes to allow companies to lease land beyond ceiling limits for plantations.

Source: Saigal, 2002

- Using forests to grow raw material for industry will be setting the clock back to the 1960s, showing that we have learned nothing from the mistakes of the past 30 years of trying to create man-made forests, which were ecological disasters, as well as completely alienating the people and leading to faster degradation.
- There are no degraded forests big enough (e.g. 2000 hectares in size) to enable economies of scale.
- The soil quality (depths of at least 1m) demanded by industry is available in India only in the best forests or farms. If good forests are not to be used for industrial plantations, the industry will have no option but to establish contacts with farmers. Even if degraded forests with soil depths of 1m do exist, they would easily regenerate on their own without much cost, if people are willing to cooperate.
- Industry has shown no interest in leasing *non-forest* wastelands when these were offered by the government; therefore, their plan to operate on equally degraded barren forest lands is highly suspect (Planning Commission, 1998).

While the working group report has significantly reduced the chances of the immediate leasing of degraded state forest land or plantations to companies, the debate has not died down.

Potential for strengthening private-sector contribution on state forest lands

Forest fringe communities

JFM is now a central forest management strategy, covering 14 million hectares. JFM has created space for private-sector players, such as forest fringe communities, to manage state forests, including plantations.

But many FPCs still depend upon external funding. JFM will be sustainable in the long term only if the communities can make a true profit.

Marketing is a major challenge for many FPCs. Many forest-based industries face shortages of particular forest raw materials, so there is potential to link the two profitably. FPCs might persuade industries to invest in the forests, tailoring production to industry's needs, while ensuring that subsistence and environmental considerations come first. Many NGOs oppose these partnerships, seeing them as an opportunity for industry to control community forests; so considerable safeguards would have to be incorporated (so far, corporate involvement has mainly been charitable, for public relations purposes).

The central government has started promoting forest development agencies (FDAs) – federations of FPCs registered under the 1860 Societies (Registration) Act, 1860. A National Afforestation Programme with 10.25 billion rupees to spend in 2002–2007 was launched to support FDAs' afforestation projects. As FDAs can involve up to 50 FPCs, it may be easier to negotiate agreements or contracts with industry at that level.[14]

Corporate sector

While current policy does not allow new commercial plantations on state lands or the leasing of degraded forest to industry, the possibility of leasing or joint management by the corporate and private sectors of some of the *existing commercial plantations* of FDCs could be explored.

The current management of these plantations is poor (see Box 8.1), and professional management could substantially increase productivity. Several companies have made considerable progress in research and development for tree improvement, especially of commercially important species such as eucalyptus, poplar and casuarina. ITC Bhadrachalam Paperboards Limited has developed eucalyptus clones that produce 20–44 cubic metres per hectare per year under rain-fed conditions; up to 50 tonnes per hectare per year has been reported under irrigated conditions. Replacing existing plantations with improved species could substantially increase yield.

Conclusions and ways forward

The Indian forestry sector's major challenges are the degradation of forests, the demand–supply gap and inadequate investment in the sector. Private-sector involvement could supplement the government's efforts to address all of these challenges.

Forest-fringe communities are already working with the forest departments under the Joint Forest Management programme to protect and manage18 per cent of forest lands, including plantations. While JFM has helped to regenerate degraded forests and plantations and to improve livelihood opportunities for the forest fringe communities, problems remain. The most important of these is the need to make the legal basis of the programme firmer so that communities' rights with respect to JFM forests are ensured. Currently, FDs can unilaterally change these and even stop Joint Forest Management after communities have invested years of effort in protecting and regenerating 'their' forests. Furthermore, as nearly one fifth of state forest land is already under JFM and this is likely to increase, the potential for commercial production from these forests and linkages between the JFM village forest protection committees and forest-based industries should also be explored.

Current policy and law does not encourage corporate private-sector involvement on state forest lands. However, considering the unsatisfactory performance of most forest development corporations, involving the corporate sector in the management of *existing commercial plantations* with the FDCs should be explored. Improving productivity through more professional management would help to ease the raw material shortages facing industry. However, the corporate sector should pay commercial rates for land and detailed guidelines should be prepared to ensure that local communities' interests are not compromised and that adequate environmental standards are maintained.

It is hoped that through these interventions both forest fringe communities and the corporate sector can contribute significantly towards the improved management of state forests and plantations.

Corporatization, Commercialization and Privatization: New Zealand

Jacki Schirmer and Michael Roche

New Zealand's forests and plantations in 2003: An overview

New Zealand's indigenous forests

New Zealand's 6.4 million hectares of indigenous forests make up just under 25 per cent of the country's land cover. Approximately 77 per cent are under Crown ownership, managed by the Department of Conservation. Of the privately owned estate, approximately 31 per cent is in Maori ownership (MAF, 2002a).

Under The Forests Act, as amended in 1993, landowners must have a sustainable management plan or permit to harvest or mill indigenous timber – although there are a few exemptions.[1] Currently, very little logging takes place in indigenous forests, with most wood supply coming from plantations.

New Zealand's plantations and plantation industry

The sections below give a picture of the plantation estate and processing industry in 2001, ten years after the first round of sales of state-owned plantations occurred.

The plantation estate
In April 2001, the New Zealand plantation estate was estimated at 1.799 million hectares (NZFOA, 2002). Approximately 71 per cent of the total area is on the North Island and 29 per cent on the South Island (MAF, 2001). *Pinus radiata* makes up approximately 89.4 per cent of the total plantation area, and

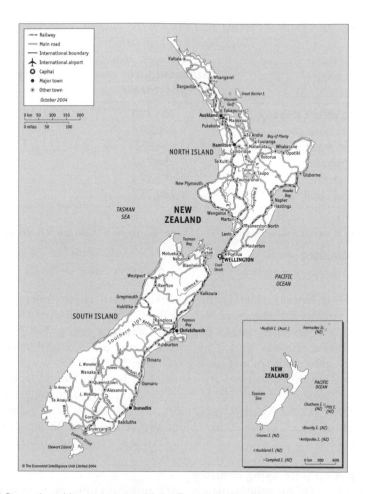

Source: © Reproduced by permission of the Economist Intelligence Unit

Figure 9.1 *New Zealand*

Douglas fir 5.7 per cent, with a number of other softwood and hardwood species making up the remaining 4.9 per cent (NZFOA, 2002).

Most plantations are currently, and were historically, established on rolling to steep hill country, with little establishment on the most productive farmland due to its value for agricultural production (MAF, 2002a).

Plantation ownership

This section examines ownership of the plantation estate in 2001 as shown in Table 9.1.

Rights to grow and harvest trees were privately owned on all but 9 per cent of the New Zealand plantation estate in April 2001. Sixty per cent was owned by 14 organizations, each owning over 20,000ha. The largest four owners

Table 9.1 *Stocked planted production forest area in New Zealand in 1 April 2001 by ownership*

Plantation owner*	Estimated total area (hectares)	Percentage of estimated total
Registered public company	780,790	43
Registered private company	868,284	48
State-owned enterprise	44,937	3
Local government	56,753	3
Central government	47,993	3
New Zealand total	1,798,757	100

Note:
*Ownership is based on ownership of the plantation trees, not on ownership of the land.

Source: NZFOA, 2002

– Carter Holt Harvey (CHH), the Central North Island Forest, Fletcher Challenge Forests and Rayonier New Zealand – each own over 100,000ha of plantations, and together own 38 per cent of the total estate (NZFOA, 2002). The rest of the estate is owned by smaller companies, partnerships, small landowners, joint ventures and local government.

The land upon which trees are established is often owned by a different party than the owner of the trees. In New Zealand, approximately 15 per cent of plantations are on Maori-owned land. Others are on land owned by the state, much of which is subject to claims under the 1975 Treaty of Waitangi Act.

The plantation processing industry

Of a total harvest of approximately 18.5 million cubic metres of wood in 2001, about 12.6 million cubic metres were processed domestically. Only about 0.03 per cent of the total volume of wood produced was sourced from indigenous forests. Approximately 13 million cubic metres were exported in raw and processed form, bringing in NZ$3.7 billion to the domestic economy (NZFOA, 2002). New Zealand's domestic processing industry in 2001 included (MAF, 2002a):[2]

- 4 pulp and paper companies;
- 6 panel board companies;
- around 350 mostly relatively small saw mills;
- 80 re-manufacturers.

According to some estimates, the domestic forestry industry has enough plantation resources to absorb NZ$3 billion more in investment in wood processing capacity, based on the assumption that 25–30 per cent of logs are exported (MAF, 2002a).

Contribution of plantations to the economy

The contribution of the predominantly plantation-based forest industries to the national economy in 2001–2002 was estimated at 4 per cent of national gross domestic product (GDP) (NZFOA, 2002) employing 24,315 people in February 2001 (NZFOA, 2002).

Regulation, legislation and policy

Domestic legislation and policy

Resource Management Act

The 1991 Resource Management Act (RMA) is the primary legislation that regulates plantation management. It applies to all resource management in New Zealand and uses an effects-based, rather than prescription-based, approach.

The RMA is administered by regional and district councils through regional and district plans. District councils have powers to control the use of land for particular purposes. Where forestry is classed as a 'discretionary' or 'non-complying' activity, plantation managers have to apply for a 'resource consent' before being allowed to undertake plantation activities. Where it is zoned a 'controlled' activity, the district council may impose conditions on the activity but cannot refuse it. In many cases, public notification is required before a resource consent can be granted. The act also includes a duty of consultation with Maori groups.

Forestry right

In order to enable effective joint ventures to take place between investors/capital providers and landholders, the 1983 Forestry Rights Registration Act allows a landowner to grant another person or entity rights to establish, maintain and/or harvest a crop of trees (Forestry Joint Venture Working Group, 1991).

Taxation

The current forestry taxation regime has been in place since 1991. Under it, some forestry costs can be deducted the year they are incurred – including some types of land preparation, road maintenance expenditure, and tree-planting and tree maintenance costs. Other costs must be held over to be deducted from the income received when the trees are harvested (Campbell et al, 2001).

Voluntary agreements and self-regulation

The New Zealand Forest Accord was signed in 1991 between representative industry groups and environmental non-governmental organizations (ENGOs). The accord specifically views plantations as a way of producing wood products while protecting indigenous forest, and identifies land on which it is inappropriate to establish plantations, particularly land supporting naturally regenerating or mature indigenous forest (Walker et al, 2000).

In December 1995, the plantation industry and ENGOs signed the *Principles for Commercial Plantation Forest Management in New Zealand*. The agreement is complementary to the Forest Accord and aims to achieve 'environmental excellence in plantation forest management' (*Principles for Commercial Plantation Forest Management in New Zealand, 1995*).

The 1990 (revised 1993) Logging Industry Research Organization's (LIRO) *New Zealand Forest Code of Practice* provides a means 'of ensuring safe and efficient forest operations that meet the requirements of sound and practical environmental management (Visser, 1996; LIRO, 1993, cited in Walker et al, 2000).

Certification

During recent years, various forms of certification have been obtained by different plantation companies and businesses in New Zealand, including the ISO 14000 Environmental Management System, the Forest Stewardship Council (FSC) and the Verification of Environmental Performance Scheme (Walker et al, 2000; MAF 2002a).

In 2002, a draft New Zealand forestry standard for plantations was developed using the FSC standards process, and is still under scrutiny (Inwood News Services, 2002a New Zealand Forestry, 2004). Development has also begun on an indigenous forest standard.

Plantation development in New Zealand: A brief history

Rates of plantation establishment

Figure 9.2 shows plantation establishment rates over time in New Zealand. The overall trend has been for increasing rates of plantation establishment, with peaks in planting occurring during the 1920s to 1930s (via the early bond afforestation companies and the state); the 1970s (as the state and large utilization companies expanded their estates); and the mid to late 1990s (driven by small landowners and investors).

State-owned plantations: Development and evolution, 1800s–1984

Early state involvement in plantation development

Prior to 1897, the majority of tree planting in New Zealand was undertaken by private landholders (Poole, 1969). Increasing concern about the potential for a future timber famine, due to the rapidly decreasing indigenous forest resources in New Zealand, led to the recommendation in 1896 that the state undertake tree planting (Simpson, 1973). In 1897, a forestry branch of the

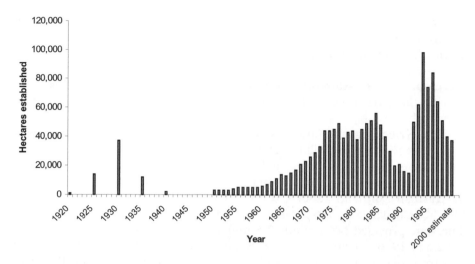

Source: MAF 2001

Note: Only five year average annual estimates are available before 1950; the 2000 planting rate is an estimate

Figure 9.2 *Establishment of new plantations in New Zealand, 1920–2000*

Lands Department was formed to undertake this role, and over the next 22 years planted 13,660ha acres of plantations.

The first planting boom of the 1920s–1930s
Various attempts were made to pass a forests act that would give adequate state control over indigenous forest management and development of state plantations, though none resulted in large-scale state involvement in afforestation until the establishment, in 1919, of the State Forest Service – later to become the New Zealand Forest Service (NZFS)[3] – which was to be responsible for all of the Crown's plantation and indigenous forest management until April 1987 (Clarke, 1999).

The newly formed NZFS carried out an inventory of indigenous forests and estimated that they would be cut out by 1965–1970, based on estimates of indigenous forest growth rates and projected future wood demand. The solution to the foreseen shortage of wood was deemed to be establishment of plantations of exotic tree species (Cox et al, 1993).

In 1924, the government announced a new afforestation strategy 'to lift the area of State plantations from 13,000 acres[4] to 300,000 acres[5] by 1935' (Kirkland and Berg, 1997, p49). Planting proceeded rapidly and then effectively stopped during the late 1930s after the goal was achieved (Kirkland and Berg, 1997).

A considerable proportion of planting in this first planting boom, both public and private, took place in the central North Island. However, state plantations were also established over other parts of both North and South Island, as part of the NZFS policy of ensuring that each province could eventually have an adequate timber supply to meet its needs.

Development of a state plantation processing industry, 1940s–1960s

During this period, new afforestation occurred at a much slower rate than previously.

The NZFS began to undertake its own milling operations in 1940, establishing the Waipa State Mill in the central North Island region because of the large area of plantations there (Poole, 1969). The NZFS wanted to be involved in developing an integrated saw mill and pulp and paper plant, and in 1952 the Tasman Pulp and Paper Group, principally formed by Fletcher (later Fletcher Challenge Forests) was developed as a joint state–private enterprise. The state provided NZ Pounds 7.5 million of advances to Tasman between 1952 and 1956 to help cover costs (Simpson, 1973) and the NZFS provided logs from state plantations at very low prices to feed the operation. The state sold its direct interests in Tasman in 1979; but its long-term supply contracts remained in place, effectively subsidizing Tasman's processing costs (Kirkland and Berg, 1997).

Plantation expansion, central planning and multiple objectives: The 1960s–1980s

The second planting boom took place from the 1960s to 1980s, and during much of this time was heavily influenced by the central planning policies of the government. Planting targets were set for the public and private sector at Forestry Development Conferences and were subsequently approved by government (Poole, 1969; Le Heron and Roche, 1985; Birchfield and Grant, 1993; Kirkland and Berg, 1997).

In 1960 the director general of forests called for 5665ha to be planted annually by both state and private agencies (Le Heron and Roche, 1985). Even higher targets were set until 1981, when the Forestry Development Conference presented targets of 43,800ha per year to be established over 1981–1985, with 16,050ha to come from state establishment, 17,200ha from large private growers and 10,550ha from small private growers.

The new planting was driven principally by a perceived need to plant enough plantations to meet future domestic demand, and the desire to develop a strong forest products export industry (Kirkland and Berg, 1997).

State assistance to the private sector

In 1976 the official objectives of the NZFS were changed to achieve varied objectives, including assisting private sector afforestation, meeting employment and regional development objectives, and meeting environmental demands.

The state put in place various incentives to encourage planting by the private sector during the second planting boom. Some incentives aimed to encourage planting on agricultural land,[6] rather than on land marginal for agriculture, which was often far from markets. Taxation and legislation reform provided incentives for this, and planting loans were provided through the 1962 Farm Forestry Act.

The majority of the area of new planting in the second planting boom was established by the private sector. However, there are estimates that state funding accounted for 70–75 per cent of total afforestation costs from 1970 to the mid 1980s (Horgan, 1990, cited in Cox et al, 1993).

Regional development and employment

The government used the NZFS to try to achieve social objectives, particularly reducing rural unemployment. This led to significant overstaffing of the NZFS, especially in many economically depressed rural regions. At the same time, claims of social and economic benefits from plantations were challenged by some members of rural communities who believed that plantations had negative impacts on agricultural enterprises and rural social structures (Aldwell, 1984; Le Heron and Roche, 1985; Roche, 1990b).

Meeting environmental demands

The NZFS managed both indigenous forests and plantations. Concern about the sustainability of indigenous forest logging, and about conversion of indigenous forest to plantations, was expressed by a growing environmental movement during the time of the second planting boom. ENGOs were highly critical of the NZFS, and called for separation of its conservation and commercial roles as a way of making it more accountable for its environmental impacts.

Final days of the NZFS

The NZFS had economic objectives to meet; yet, it had not managed to turn an operating surplus, requiring government support to fund its operations. At the same time, it was effectively subsidizing wood processors by selling its logs at low prices (Birchfield and Grant, 1993).

The NZFS was a clear candidate for some kind of change. The 1984 Labour government budget increased the prices of logs from state plantations and forests and removed forestry investment concessions (Birchfield and Grant, 1993, p25). In 1985, it was announced that the commercial and non-commercial functions of the NZFS would be separated.

Privately owned plantations: Development and evolution, 1800s–1980s

Private planting, 1800s–1920s

The first direct incentives for private planting came with the passing of the 1871 Forest Trees Planting Encouragement Act; but this achieved only limited success (Roche, 1987, p49). Most private planting was undertaken without recourse to assistance schemes, and was strongly influenced by the belief that trees could affect climate and rehabilitate wasteland (Roche, 1987).

The bond holding companies: Private planting during the 1920s–1930s
In the private sector, planting during the first boom in the 1920–1930s was undertaken primarily by companies selling bonds entitling the owner to an acre of land upon which the company would establish trees and maintain them for a length of time (Roche, 1990a).

However, bond companies came under considerable criticism for the methods they used to convince people to invest, the profiteering of some companies and their plantation establishment methods. The consequent establishment of a commission of inquiry into company promotion methods in 1934 led to the end of the bond-selling era (Roche, 1990a).

The second planting boom: The private sector, 1960s–1980s
The private plantings of the 1960s were largely undertaken by large forest utilization companies, such as New Zealand Forest Products (NZFP)[7] and the Tasman Pulp and Paper Company, which aimed to supply their own mills with wood (Le Heron and Roche, 1985). NZFP, for example, decided to expand its plantation estate to 161,871ha in 1969 and 202,339ha in 1973 (Le Heron and Roche, 1985) in order to supply its future processing needs.

Revision of the taxation system in 1965 allowed companies to deduct various costs of establishment and maintenance of plantations from current income, thereby contributing to the expansion of planting (Le Heron and Roche, 1985, p220). However, after 1984, when many incentives for planting were removed (Cox et al, 1993), total new establishment rates (public and private) dropped from around 56,000ha in 1984 to 16,000ha in 1991 (Mead, 1995).

Corporatization and privatization of state plantations: 1984–2003

The 1984 Labour government was elected following a severe economic crisis (Birchfield and Grant, 1993). It inherited a sluggish domestic economy, rising foreign debt and high unemployment, and focused on addressing the economic problems facing the country (OECD, 1999) through a structural adjustment programme, part of which aimed to reduce 'the size, role and power of the state through corporatization, privatization, devolution and managerialism'(Kelsey, 2002).

Corporatization

The decision to corporatize
In 1985, the government announced that the NZFS would be corporatized. This decision was driven by two primary forces. The first was the government's overall structural adjustment programme, in which commercial and regulatory functions of government were to be separated by creating state-owned enterprises (SOEs) and subsequent privatization of many of these (Clarke, 1999). The second was pressure from ENGOs to have the conservation and wood

production roles of the NZFS separated, and a new department created specifically to be responsible for protection of the environment (Birchfield and Grant, 1993).

The process of corporatization

In February 1986, the Forestry Corporation Establishment Board was established 'to advise on the form and function of [the new] organization' (Roche, 1990b, p951). The board was free of vested interests in forestry from either the public or private sector.

The board recommended full privatization; but the government decided instead to corporatize the commercial functions of the NZFS (Roche, 1990a, 1990b). The New Zealand Forestry Corporation (NZFC), a limited liability company, was established to manage all commercial forestry operations on 1 April 1987 (Clarke, 1999). The NZFC had one principle objective: to operate as a profitable business. The social and environmental objectives that the NZFS formerly had had to meet were now the responsibility of the Department of Conservation (DOC) and the Ministry of Forestry (MOF).

Perhaps the biggest issue in the transition to corporatization was employment. The NZFS had been considerably overstaffed; the NZFC had a mandate to operate as an efficient business, and needed to operate with a smaller number of employees than the NZFS had. The methods used to achieve this needed to ensure that the industry would function effectively.

Early on, the board decided to re-employ as many NZFS employees as possible in the new NZFC, but without automatic transfer of staff, enabling it to create a new employment structure more suited to a commercial corporation – with less management layers, encouragement of competition between managers and elimination of many jobs (Birchfield and Grant, 1993).[8] The way in which this change was achieved differed for salary and wage workers.

For salaried staff a *Green Book* was produced which listed the positions available in the NZFC. NZFS employees had to transfer to DOC or MOF, apply for jobs in the NZFC, or take a redundancy package offered by the government. Managerial and staff positions in the NZFC were negotiated on individual contracts.

Wage workers were largely eliminated in the NZFC, which instead engaged a much lower number of independent contractors. A series of discussions were held between the timber workers' and workers' unions, the NZFS and the State Services Commission, resulting in agreement on a process for deciding which wage workers stayed and which left.[9] At the same time, the NZFC offered an Enterprise Opportunity Scheme that helped arrange finance for those who wanted to become contractors to purchase equipment and set up as private contracting business entities.

There were no compulsory redundancies; but a considerable number took voluntary redundancies, and there is evidence that many did not find work for a long time afterwards and/or had to shift to different regions in order to find work (Birchfield and Grant, 1993).

Changes to employment, log pricing and business structure resulted in the NZFC making a NZ$53 million profit in its first year of operations (Roche, 1990b) and generating NZ$174 million in cash surpluses, compared to a NZ$117 million deficit on commercial activities in the final two years of the NZFS (Birchfield and Grant, 1993). The NZFC's success was short lived, however, with the government deciding to privatize the corporation.

Changes accompanying corporatization

In 1984, long-term contracts were held over approximately two-thirds of the high quality logs coming from state forests; but these only brought in around half the revenue received by the NZFS. The realigning of long-term contracts with CHH and Tasman with market values, discussed further below, along with a shift to tendering for short-term contracts, began to bring in higher prices for timber from the late 1980s (Birchfield and Grant, 1993; Kirkland and Berg, 1997).

The removal of various import controls and export subsidies forced rapid change and restructuring in the processing industry and resulted in more exposure to global economic fluctuations (Grebner and Amacher, 2000; Stringer, 2002). At the same time, deregulation of the transport industry, the ports and the labour market meant that processors faced lower costs for transport of logs and products and were able to have a more flexible labour force, allowing efficiency improvements.

Privatization

The decision to privatize

In 1988, the government announced that plantations would be privatized (Clarke, 1999), saying that their primary motive for privatization was to use the sale to reduce public debt (Roche, 1990b). In reality, a range of factors motivated the decision to privatize, including the government's broader economic ideology of privatizing all SOEs unless there were compelling economic or social reasons to retain them in government ownership (Clarke, 1999); concerns that 'the SOE structure might only incorporate the worst of the public and private sectors' (Roche, 1990b, p952); concerns that the NZFC would be subject to political interference (Clarke, 1999), which helped to push the privatization process; concerns about the NZFC's limited capacity to raise capital for further investment in expanding its operations (Birchfield and Grant, 1993; NZIER, 2000); and a belief that major processors would achieve security of wood supply if state plantations were privatized.

Privatization was also seen as a way of resolving an ongoing disagreement between the Treasury and the NZFC over the value of the NZFC's assets.[10]

Deciding the form of privatization

After the decision was made to privatize, the Forestry Working Group – which included representatives from government and the private sector – was formed to make recommendations about how privatization should take place.

The decision had already been made to sell NZFC's assets, rather than to sell the NZFC as a going concern, primarily as 'it would take a considerable time to rationalize the business to the point of it being a credible contender for public listing' (Birchfield and Grant, 1993, pp12–13).

One of the first decisions to be made was whether both the trees and land should be sold, or only the trees. Ongoing claims over large areas of land under the Treaty of Waitangi meant that selling government land was unlikely to be acceptable. The decision was made to sell the trees but not the land, through selling long-term cutting and management rights in the form of long-term leases called Crown Forestry Licences (CFLs).

Sale of the trees

Designing the sale of the trees

Size of plantation units to be sold

Under the 1989 Crown Forest Assets Act, the plantations were divided into 90 units appropriate for plantation management and harvesting. The units varied in size from 51ha to 132,112ha, in theory ensuring maximum flexibility by allowing both small and large organizations to bid for units, and hence maximizing the total potential sale price of the assets (Clarke, 1999).

Covenants and restrictions on the Crown Forestry Licences

Each unit had a CFL assigned to it, with 'individual terms and conditions of sale' (Clarke, 1999, p37). The CFLs were designed to be as close to freehold land with rent as possible in order to minimize the constraints placed on CFL holders. Many plantations had not been adequately surveyed, and had to have rights held over them identified and written into the CFLs as covenants (e.g. provisions for vehicular access rights, conservation requirements and grazing rights). All CFLs included a 'wander at will' clause allowing pedestrian access to the plantation unless safety or other reasons precluded access.

During the initial round of sales in 1990, no replanting conditions were imposed. By the time of the 1992 and subsequent sales, a condition was imposed that the land must be replanted or converted to another sustainable use that had to be approved by government.

Similarly, bidders were not initially required to guarantee to maintain or expand domestic processing; but by 1996 'the sales process required potential bidders first to demonstrate their intention to add value to the resource within [New Zealand]' (Clarke, 1999, p43).

All CFLs could be terminated in the event of land being transferred to Maori owners. Once a termination notice was given, the land would be returned as the trees were harvested.

Selling the Crown Forestry Licences

A competitive sealed bidding system was used to sell CFLs. The primary determinant of successful bids was price. Bidders could bid for any combination of the 90 plantation units they chose. The aim was to allow as wide a range of potential bidders, both large and small, as possible into the sale process (Birchfield and Grant, 1993).

Pressures and issues during the bidding process

Concerns over the potential for a monopoly situation arising if Tasman Forestry (Fletcher Challenge), Elders–NZFP or Carter Holt Harvey purchased a large proportion of the plantations led to a decision by the Commerce Commission to forbid each of these organizations from buying more than a specified area of plantations in particular geographic regions.

There was considerable pressure to proceed with the sales as rapidly as possible in order to resolve supply issues for the processing industry. While the privatization process was being developed and undertaken, processors were given maximum one-year contracts for wood supply from NZFC (Birchfield and Grant, 1993).

Existing contractual obligations

Long-term contracts with both Tasman Forestry and with Carter Holt Harvey had to be dealt with before privatization of the plantations affected by the contracts could go ahead.

The NZFS had a log supply agreement with Tasman; but in the contracts, according to Birchfield and Grant (1993, p183), 'the question of price adjustments over time was quite superficially addressed'.

When the government increased log prices in 1984, drawn-out negotiations with Tasman failed and the plantations that were required to supply logs under the agreements were taken out of the sale process and, instead, transferred to a new Crown subsidiary, New Zealand Timberlands·(Bay of Plenty) Ltd (Birchfield and Grant, 1993).

Carter Holt Harvey also had long-term supply agreements with NZFS and, following a rejected bid for the plantations in question, took court action against the state, claiming assurances of long-term wood supply were not being honoured. The court action was not concluded as an agreement to sell the plantations to CHH was made before a decision was handed down in the case (Birchfield and Grant, 1993).

The sales

The first round of sales

In April 1990, the NZFC called for tenders for 66 of the 90 units (Clarke, 1999). When the bidding closed, the total of the bids was considerably lower than 'a previously calculated expected market value' (EMV) (Birchfield and Grant, 1993, p228).

'It was decided that a Treasury-led negotiating team should approach two bidders whose tenders were within 90 per cent of estimated market value. Both were prepared to negotiate and restructure their bids' (Birchfield and Grant, 1993, p229). The first two sales were announced in July, with 73,000ha of plantations sold for NZ$364 million to the two successful bidders, Fletcher Challenge and Ernslaw One Ltd.

Several further sales occurred in 1990 after negotiations by the Treasury with other bidders (Kirkland and Berg, 1997). In total, CFLs to over 250,000ha of plantations and other assets, representing 44.7 per cent of the plantations and forests offered for sale, were sold in 1990 for over NZ$1 billion (Birchfield and Grant, 1993; MAF, 2002a).

The unsold plantations

The unsold plantations were formed into an SOE, New Zealand Timberlands Ltd. Many of these plantations had attracted low bids due to doubt over their commercial viability, so the initial focus of New Zealand Timberlands Ltd was on setting up markets and sales contracts to prove that the plantations had commercial value.

There were now three SOEs managing the assets formerly managed by NZFC. Two – Timberlands (Bay of Plenty) and New Zealand Timberlands Ltd – managed plantations, and one – Timberlands (West Coast) – managed indigenous forests on the West Coast.

Subsequent sales

The sale of New Zealand Timberlands Ltd's 100,000ha to ITT Rayonier for NZ$364 million was achieved in 1992 (Clarke, 1999; MAF, 2002a). Timberlands Bay of Plenty[11] was sold to a consortium led by Fletcher Challenge for NZ$2026 million in 1996 (Clarke, 1999; MAF, 2002a).

Plantations retained in Crown ownership

The Crown had established around 65,000ha on land leased from Maori owners from the late 1960s to 1970s (Schell, personal communication). The length of lease ranged from 33 to 99 years. The leases were developed for a variety of reasons, including erosion control and catchment protection, and regional development objectives, with commercial return from the trees not always a priority. All of the leases had one thing in common: they contained no right of assignment. The Crown could not transfer its interest in the lease to another party.

The Crown wanted to divest itself of its interest in the leases, but was effectively dealing with a 'one buyer, one seller' situation. The policy of the Crown has been that, where possible, it will divest itself of its interest in the leases, but only if the outcome of doing this for the Crown is either neutral or positive in an economic sense. The Crown has divested itself of some leases, but still manages others.

Two mechanisms have been used to divest the leases to date: outright sale and surrendering the lease.

Outright sale

Outright sale has involved selling the Crown's interest in the lease to the Maori landowner, or by an organized back-to-back sale in which a purchaser is found for the plantation, and the Crown's and Maori landowner's interests are sold on to the purchaser.

Surrendering the lease

Many Maori landowners, however, have not had the financial resources to directly buy out the Crown's interest in the lease. As a result, on the majority of the lease land, the Crown is exiting the leases by surrendering the lease after one rotation. This is done for stumpage sharing leases, and involves recalculating the percentage return to the partners. The recalculation must enable the Crown to break even in net present value terms with the expected returns it would have received had it operated for the entire length of the lease. The recalculation must also ensure the landowner has the resources to manage the plantation once it is returned to them. Once agreement is reached, the land is handed back as the trees are harvested.

The Lake Taupo Forest Trust (LTFT) provides an example of this process, being the single largest Crown lease on Maori land for plantations. When the two parties agreed to negotiate a lease surrender after one rotation, they decided to each calculate a stumpage share that would meet their needs. For LTFT, the needs were to be able to manage the plantation and to pay dividends to their 10,000 owners from harvest returns. The Crown, meanwhile, undertook its own calculations to find a stumpage share that would give it a return equal to the estimated net present value of the benefit it would have received had it stayed in the lease until it ended in 2039. The two parties then negotiated until they settled on a share of 35 per cent of stumpage to the LTFT and 65 per cent to the Crown.

Land issues: Ownership, rent and transfer

Land claims and privatization: Ensuring Maori rights were preserved over land

Maori land claims were lodged over much of the land on which state plantations were established, and had yet to be resolved by the Waitangi Tribunal.

The 1986 State owned Enterprises Act required that the Crown's responsibilities under the Treaty of Waitangi be preserved over any land transferred to the control of SOEs. The Forestry Working Group had to find a way of fulfilling these obligations in a way that resulted in maximum value to the state from the sale (Birchfield and Grant, 1993).

The Forestry Working Group believed that splitting ownership of the trees and the land and selling only the trees would resolve the issues. However, there was concern from the Maori Council that the proposal, in which cutting rights to two crop rotations would be sold, would prevent the Maori from using their land for up to 50 to 70 years after land claims were resolved in favour of the Maori.

Subsequently, an agreement was formalized in which, when the trees were sold, the state would 'charge [the purchaser of the trees] a land rent and ... hold proceeds in trust[12] for whomever the Waitangi Tribunal might rule to be the ultimate owner of the land' (Clarke, 1999, p39).

This and additional conditions[13] were written into the Crown Forestry Licence.

Determining levels of rent

There is a periodic rent review for land underlying CFLs every third year, and a general review every ninth year. Rent was originally calculated at 7 per cent of land value but has been disputed between some CFL holders and the Crown, with some disputes being taken to arbitration. Consequently, the Crown switched to using a market rental rate to determine part or all of the rent for many CFL holders. Debate over appropriate methods of determining rent is ongoing.

A Crown review found that delays in obtaining agreement in the rental review process were not so much over the value of comparable land in the private sector, but rather over the adjustments to be made to reflect the covenants placed on different CFLs.[14] Another problem was that new people were often negotiating on behalf of the CFL holder every three years; as a result, agreements on basic facts achieved in previous agreements had to be worked through again. The Crown began implementing a system of developing protocols with CFL holders in which both parties agree on the set of conditions applying to a licence in order to achieve consent on the underlying information on which the valuation is to take place.

Transfer of land to Maori owners

Since sale of the CFLs, there have been some transfers of land ownership from the Crown to Maori claimants, although the majority of claims have yet to be settled. It had been expected that the Treaty of Waitangi Tribunal would settle individual claims by making a binding recommendation under which land would be transferred from the Crown to Maori owners.

However, the settlement process has generally taken a different form. The Crown has been settling groups of claims in a single cash settlement, and has then been offering Crown land for purchase by the group that has received that settlement. When this occurs, and land covered by a CFL has been purchased from the Crown by Maori, the conditions in the CFL relating to termination of the CFL and return of land to Maori owners apply, and the Maori owners receive the funds that have been held, to that date, in the Crown Forestry Rental Trust.

While this process has operated relatively smoothly, there have been particular issues for Maori looking to purchase land covered by CFLs – in particular, getting agreement on valuation of the land and the difficulty in obtaining finance.

The use of a cash settlement followed by land purchase requires the parties to agree on a valuation of the land being purchased. Various valuation issues have already been discussed above, and the same issues apply to this process.

Maori groups may wish to borrow money to be able to purchase land offered to them under the settlement over and above the amount received in the cash settlement, and/or to allow them to finance management of plantations once land is returned to them. Either of these cases involves obtaining finance, which is difficult without the guaranteed income previously provided by an annual rental income from leaseholders.

As a result of this, groups such as Ngai Tahu (a Maori iwi) in the South Island have developed their own leases, based on the Crown Forestry Licence, and negotiated with the companies holding the CFLs to develop a lease agreement once land has been transferred to the Maori owners. This means income continues to be received from rent, giving a return from the land to the owners and making it easier to obtain finance.

Many in the plantation industry are concerned that some proposed Maori land transfers, which include roads providing access to other plantation areas, could impose significant costs on the industry. This points to the importance of making clear provision for access rights and road use in plantation areas in the privatization process, particularly where trees and land are separated.

Post-privatization: The 1990s planting boom

The rise of farm forestry and small owners
During the early 1990s, the third planting boom began with an estimated 85 per cent of planting undertaken by small plantation owners rather than large companies (Mead, 1995).

This boom was driven partly by the 1991 reintroduction of the deductibility of tree-planting and maintenance expenses from current income (Cox et al, 1993; Nagashima et al, 2002) and by increases in international saw-log prices in 1993 (Ministry of Forestry, 1993).

Post-privatization role of government in plantations
Despite the almost complete cessation of direct commercial involvement in the industry, government still plays a range of roles that influence and often assist the development of the plantation industry. These roles include (Walker et al, 2000):

*   Ensuring a competitive trading environment for industry, including facilitating market access, supporting industry initiatives and providing adequate infrastructure.[15]
*   Disseminating information on the environment and on technology through the Ministry of Agriculture and Forestry.
*   Providing funding for the East Coast Forestry Project.[16]

Evolving demands and roles for plantations

For both the public and private sector, motivations for plantation establishment have changed over time, from avoiding a future timber famine and improving climate, to providing a domestic timber supply, and finally to providing timber for an export industry. Continuing demands for improved environmental performance have been made on both sectors. The state, however, has had to meet a range of other demands which the private sector has not, including:

- Use of afforestation to reduce rural unemployment and provide opportunities for regional development.
- Planting for erosion control and environmental improvement.
- Increasing demand for multiple-use management for recreation, aesthetics and environment.

Balancing acts: Reconciling public-policy objectives and private-sector investment

The following section compares the outcomes under the NZFS with the outcomes of privatized plantations.

Economic outcomes

Several positive economic outcomes of privatization are often claimed. These include the following (Ministry of Forestry, 1993; MAF, 2002a):

- development of a more competitive industry;
- increased resource security and certainty, resulting in investment in processing;
- the creation of a secondary market for plantations that had not existed previously;
- better access to overseas markets and distribution;
- the updating of the saw-milling industry since it was forced to compete in a more open market environment.

Many of those interviewed for this study believed that privatization has led to increased efficiency and competitiveness, particularly in global markets. However, many also believed that it was not possible to separate the impacts of market, labour, port and transport deregulation from the impacts of corporatization and privatization.

Other factors deemed responsible were the period of worldwide economic growth that coincided with privatization; 'the freeing of government businesses from social obligations, such as job creation' (Easton, 1994, cited in Hall, 1997, p185); and the log price increase of 1993 (Clarke, 1999).

Conversely, Grebner and Amacher (2000) found that deregulation and privatization had negative impacts on cost efficiency in the short term, although they made no conclusions about the long-term effects.

One of the goals of privatization was to increase security of wood supply in order to encourage investment in domestic processing. Clarke (1999, p43) found that 'Of the NZ$1600 million of intended investments [in value-added processing] in the period 1990 to 2005, 90 per cent is attributable to the purchases of state forest assets.'

However, some question whether investment in processing may, in fact, have been less than would have occurred under corporatization, as privatization meant capital had to be tied up in buying trees, leaving less available to invest in processing facilities. In addition, some believe it is possible that the same economic outcomes overall could have been achieved through corporatization without subsequent privatization, as both the NZFC, and subsequent plantation SOEs, ran their businesses profitably in competition with the private sector.

Social impacts: Employment

There have been concerns that corporatization and privatization had negative social impacts on small rural communities, primarily through changing job opportunities in rural areas.

Any study of impacts needs to examine how different sectors of rural communities have fared (Scott et al, 2000). Roche (1990b, p951) pointed out that:

> *Of the NZ$105 million that the government paid out in voluntary severance to public servants, NZ$65.7 million went to employees of the Forest Service. The social impact was also concentrated in spatial, class and ethnic terms in that the predominantly Maori workforces of some central North Island timber towns such as Kaingaroa were virtually all made redundant.*

Most agree that the impacts of redundancy were severe, particularly in small rural communities where many ex-NZFS workers were unable to find work after being made redundant. However, many believe this process was inevitable over time.

The debate concerning impacts on employment is not just over the number of jobs, but over the quality of remaining jobs. Pawson and Scott (1992, p384) suggest that the shift to contracting may not have resulted in improved quality of life for many workers. Critics believe the shift to contracting out many services means that many small contractors have become dependent on shifts in international markets for their livelihoods and, as a result, experience low work security and certainty.

Furthermore, there has been difficulty in obtaining enough people to work in plantation management and harvesting. Research on the East Coast of the North Island recorded that forestry work was considered difficult, poorly paid and dangerous, and that contractors could not be trusted to treat their employees fairly (Tomlinson et al, 2000).

Clearly, there are labour issues; however, the extent to which they are related to corporatization and privatization cannot be estimated here.

Social outcomes: Recreation and access

Many CFLs had covenants requiring maintenance of vehicular access or recreation-related infrastructure placed on them. However, some suggest that, while most facilities which existed at the time of privatization have been maintained, further development has not occurred and occasionally road access has been stopped where it previously existed.

Similarly, Knock (1993) found that recreation access decreased under the NZFC following corporatization, and that recreation opportunities either remained static or decreased in many plantations following privatization.

Hanmer Forest, a plantation providing a backdrop to the tourist town of Hanmer Springs on the South Island, is an example of the types of issues that can arise over recreation. The Hanmer Forest has been used for recreation for decades, and has also provided aesthetic benefits to Hanmer Springs, a tourist town, as the plantations cover the hills behind the town. Covenants were placed on the CFL requiring the CFL holder 'to preserve and protect the natural and historic resources ... and, in particular, to protect the landscape of the Hanmer Recreation Area' (Keey, 2001).

Concerns arose over the number of times that public access was stopped for safety reasons, and over the impact of harvesting on landscape and aesthetic enjoyment of the plantation. However, the primary issue appears to be one of who should pay for the maintenance of recreation trails that provide a benefit to the local and tourist community.

More concern arose as a result of the land being sold in 2000 to Ngai Tahu as part of their settlement with the Crown. The transfer out of Crown ownership meant that the 'wander at will' provision no longer applied. The covenants regarding management for recreation and landscape remain, but can be reviewed now that land has reverted to Maori ownership (Keey, 2001). While the Ngai Tahu have indicated that they are willing to allow access as long as the leaseholder managing the plantation is also willing, concerns have been raised that there are now no long-term guarantees of access to the Hanmer Forest.

Environmental outcomes

The separation of commercial and conservation roles when corporatization of plantations occurred realized a significant ENGO goal. It also benefited the plantation industry, allowing it to move on and operate without being tied up in conflict over logging and conversion of indigenous forest. Many believe that corporatization and privatization had a positive effect on environmental outcomes in plantation management because they opened dialogue and agreement between plantation owners and ENGOs. Under the NZFS, such dialogue was difficult as the issue of indigenous forest logging and conversion had to be addressed at the same time. However, some issues do remain. A key issue is the lack of control over clearance of regenerating indigenous scrub for plantation establishment by those individuals and companies who are not signatories to the *Forest Accord* or the *Principles for Commercial Plantation Forest Management in New Zealand.*

The role of certification

Some who believe that corporatization and privatization reduced the level of social and environmental benefits produced by plantation in New Zealand believe that increasing use of certification is effectively bringing social and environmental considerations back into prominence in plantation management. Whereas under the NZFS there was a government mandate to manage for multiple outcomes, under certification there is a market-driven motivation.

Other outcomes and issues

Before privatization, it was common for foresters to work first for the NZFS, and then to shift to the private sector. The NZFS effectively provided a training ground for the private sector, and privatization removed this 'hidden subsidy', according to foresters quoted in Birchfield and Grant (1993).

Corporatization and privatization have caused a shift to more processing and market-related research, and some believe that there is more focus on short-term gains, potentially at the expense of longer-term research with less direct commercial outcomes.

Concern has been expressed over the level of foreign ownership of plantations in New Zealand, with a view that foreign ownership results in profits leaving New Zealand and in negative impacts arising from increased exposure of the New Zealand economy to global market cycles. Opponents of this view point out that there has been increased investment in the domestic economy by many of these foreign companies since they purchased plantations from the state (see, for example, Horton, 1995; Swale, 2001).

Reduction of public debt was an often-stated goal of privatization, although there are contrasting beliefs as to whether privatization did (Kelsey, 2002) or did not (Hall, 1997) contribute to achieving this.

Lessons from New Zealand's experience

New Zealand's experience provides a number of useful lessons and principles to assist in the design of privatization in other countries.

Ensure an appropriate market environment. Economic success associated with corporatization and privatization in New Zealand probably depended upon the wide-ranging market restructuring that opened up international trade, removed barriers to trade, and deregulated ports, transport, labour and finance markets. It is also important to ensure the regulatory environment is appropriate and effective in regulating management of the newly privatized plantations, while providing an environment where commercial operations are not unduly constrained.

Use covenants to protect the interests of other users. When designing sales, it is important to ensure that appropriate covenants are in place on the resources sold and/or leased to protect various interests (e.g. environmental, cultural and heritage values and access to the resource). At the same time, it is important to recognize that the placing of some covenants on sales is likely to reduce the value received from those assets at sale, and so covenants must be clearly identified and specified in a way that does not restrict commercial operations more than necessary. Where plantations may be of higher value for their non-commercial benefits, consideration should be given to retaining those plantations in Crown ownership.

Establish a market value for the assets. Being able to prove the commercial value of plantations while in Crown ownership may help to increase sale value, and sales should be designed with this in mind. Shifting to corporatization and proving the ability to make a commercial return may, in New Zealand's case, have increased the eventual sales price of the trees.

Ensure that the plantation business is commercial and competitive. When the NZFS was corporatized, the NZFC operated with a completely different structure, which enabled a rapid transformation into a commercial business; had the old NZFS structure been transferred to the NZFC, it is questionable whether such rapid change would have occurred.

Corporatization may be a sufficient step. Consideration should be given to whether corporatization would achieve the desired changes and if full privatization is necessary. In particular, if a goal of privatization is to encourage domestic investment, consideration should be given to whether selling trees – and, hence, having private firms tie up capital in trees – is the best way to achieve this.

Identify negative impacts of change. It is important to identify the potential and probable negative impacts of privatization on the community, and, where possible, to develop programmes to ease the transition to privatization – for example, by implementing retraining programmes to assist people in developing their own businesses, as occurred when the NZFS was corporatized.

The role of certification. The use of certification may help to ensure that the newly privatized plantations still have incentives to meet a wide range of social and environmental imperatives, as well as commercial ones.

Separation of land and trees. The separation of ownership of trees and land allows for rights held over the land by indigenous groups to be preserved. However, it also requires careful identification of associated issues, such as the rights of landowners to impose particular conditions on the owners of the trees. Clarifying the issues of ownership and rights of access over roads is particularly important, and can be as important as clarifying ownership of trees and land.

Clear procedures for charging rent. Where ownership of land and trees is separated and the trees are privatized, the process for charging rent for use of the land must be clearly set out and designed to reduce the potential for costly litigation and arbitration over rent review processes.

Ensure the capacity of new owners. Where the land underlying plantations is transferred to indigenous owners, the processes used must provide the indigenous owners with the ability and resources to take on and manage the land – for example, through assisting in accessing finance for land management, or provision of resources to assist in other areas.

Facilitate interaction between landowner and tree owner. Appropriate mechanisms (e.g. CFLs) should be designed which allow the owners of land and trees to interact without compromising the interests of either party.

Develop clear guidelines on the ways in which government will remain involved in the sector. Clear guidelines provide greater certainty and consistency for the private sector.

Conclusions and ways forward

The importance of matching the instruments used to the needs and rights of different groups is clear from New Zealand's experience of separating ownership of trees and land, as is the need for ensuring appropriate rights and conditions of access and use. Similarly, the importance of reforms to other sectors of the economy for the successful commercial operation of privatized plantations is also clear, although, in this context, the question remains of whether privatization has been primarily responsible for the economic successes of parts of the plantation industry. This issue needs closer exploration in order to develop a better understanding of the key reforms that influence the success or otherwise of privatization.

10

Juggling Social and Economic Goals: South Africa

Maud Dlomo and Mike Pitcher

Introduction

By the time of South Africa's first democratic elections in 1994, the forestry industry was making a significant contribution to the national economy. The industry had matured quickly and included two world leaders, Mondi and Sappi. South Africa has favourable growing conditions, world-class forestry research and development (R&D), established timber-processing operations, and is well placed to exploit local and international markets. The main factors that have underpinned the industry's success have been the:

- strong local demand for forest, combined with limited overseas competition;
- availability of suitable land to permit the rapid expansion of plantations;
- supportive government policy;
- the initial direct role played by the state in creating forest resources (Chalmers, 2001).

Overview of forest resources in South Africa

Commercial plantations
South Africa has approximately 1.5 million hectares of commercial plantations, or 1.2 per cent of the total land area of 122 million hectares. Though relatively small, the plantations are generally extremely productive, occurring where growing conditions are favourable and supported by extremely high standards of silviculture and tree improvement. The plantations support a large wood-processing and manufacturing sector.

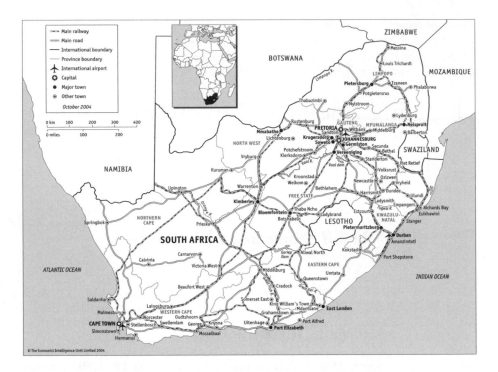

Source: © Reproduced by permission of the Economist Intelligence Unit

Figure 10.1 *South Africa*

South Africa produces 2.8 million tonnes of pulp (1.63 per cent of global supply), 2 million tonnes of paper (0.76 per cent of global supply) and 1.3 million cubic metres of sawn timber (0.3 per cent of world supply). The forestry and associated processing industries generate US$1 billion annually or 1.5 per cent of South Africa's gross domestic product (GDP).

Plantations are concentrated in a few provinces where rainfall exceeds 800mm per year, specifically in Mpumalanga, KwaZulu Natal, Eastern Cape, Western Cape and Limpopo. Pine makes up 52 per cent of the plantation estate, eucalyptus 39 per cent, wattle just over 7 per cent, with the balance comprising other species, such as poplar for match manufacturing. The historically state-owned plantations have primarily been geared to produce saw logs, whereas the privately owned plantations were mainly established to produce pulpwood.

Plantation companies are obliged to manage another 500,000ha of wetlands, indigenous forests, grasslands and infrastructure for biodiversity, watershed protection and a range of social benefits. While South Africa's plantation area continues to increase, the pace has decreased from a peak of 45,000ha per year in 1991, and since 1996 new afforestation has been around 11,000ha per year (Mayers et al, 2001).

Natural closed canopy forest

South Africa has 500,000ha of closed canopy indigenous forests – 0.3 per cent of the country. Forest area was probably always very small; nevertheless, these forests are environmentally and socially significant and are protected (Mayers et al, 2001).

Natural woodlands

Larger and more important economically, environmentally and socially are South Africa's 23 million hectares of woodlands (19 per cent of total land area – probably half its original extent). Their contribution is not well understood, but up to 20 per cent of South Africa's total energy consumption is derived from biomass (Mayers et al, 2001).

Ownership of plantation resources in South Africa

Prior to privatization in 2000, the state owned approximately 30 per cent of the 1.5 million hectares of plantations, two large companies (Sappi and Mondi) owned 47 per cent, 22 per cent was owned by smaller private companies and individual farmers, and approximately 1 per cent was owned by many thousand small growers on the 'out-grower' model (see 3.3) (Mayers et al, 2001).

Since privatization:

*   20 per cent of the state's plantations have been transferred to private-sector consortia;
*   18 per cent have been earmarked for conversion to alternative land uses, including nature conservation and land reform;
*   45 per cent is subject to disposal tender processes (as of July 2002).

It is realistic to expect that within three years all of the state's plantations will be either transferred to private-sector management or converted to other land uses.

Poverty in South Africa

Although an upper middle-income country, the majority of South Africa's 44 million people are poor or very poor and vulnerable to change.

Most of South Africa's poor live in the rural former 'homelands', where the population was concentrated during the apartheid period. There is a framework for land reform but progress is slow.

Although only 45.8 per cent of the population are rural, 70 per cent of people living below the poverty line live in rural areas. More women than men living in rural areas and female-headed households are more likely to be poor.

Most poor rural households depend upon multiple livelihood strategies, including waged employment, welfare grants, pensions, income from local enterprises and land-based activities, as well as crop production and livestock rearing. Agricultural production contributes only about 10 per cent of total rural household income for black South Africans.

Rural poverty is increasing as jobs are being lost because rural economies depend upon migrant labour remittances and there are few alternative employment opportunities (DFID, 2002).

How forestry benefits the poor

Employment
The forest and forest products industry employs over 200,000 people – about half in production forestry and half in wood processing.

More than 15,000 people are employed by contractors. The trend towards outsourcing has cost jobs and reduced wages and working conditions, but does offer an important avenue for creating black-owned enterprises in rural areas.

Direct participation in timber growing
Approximately 19,000 households are involved in small-scale commercial timber production, mostly in KwaZulu Natal Province; this is significant economically for those communities.

Direct access to forest products and other resources
Poor households living close to natural and plantation forests can access land for grazing, water supplies and a wide range of forest products, including firewood, building materials, wild foods and medicinal plants. There is also important social infrastructure associated with commercial forestry, including roads, housing and, in some cases, schools and clinics.

Participation in forest-based enterprises
Forests also support a wide range of other small-scale enterprises, including small-scale saw milling, furniture manufacturing, firewood vendors, resin tappers, and the collecting and selling of mushrooms, honey and medicinal plants (DFID, 2002).

Policy Context

Forestry
South Africa's 'White Paper on Sustainable Forest Development' was launched in 1996. It aims to promote a thriving and sustainable forest sector in all components of forestry, not just commercial forestry. The government stated its intention to withdraw from ownership and management of plantations and to assume the role of regulator and facilitator.

The National Forest Act (1998)
The 1998 National Forest Act provides a range of legal instruments through which the state can regulate the sector and transfer powers and functions for

management of state forests to the private sector, communities and other tiers of government.

Land

Through systematic dispossession, forced removals and other government abuses, black South Africans were removed from their land and relocated to the so-called 'homelands'. South Africa's 1997 land policy and legislation requires that where land changed hands through discriminatory legislation or other unjust means, such land is subject to possible restitution. Where land is held in trust by the state (as in the former homelands), the state is holding that land on behalf of communities who retain the underlying land rights.

Many of South Africa's state forests are on land held in trust by government for communities. The underlying rights to this land will be recognized in time; but the slow pace of land reform in South Africa remains a widespread cause for concern.

Water resources

The South African government estimates that the limits of economically usable, land-based freshwater resources will be reached during the first half of this century A National Water Act was promulgated in 1998 to guide responsible authorities in the issuing of water-use licences. All tree growers require licences for afforestation over 10ha (Mayers et al, 2001).

Government needs to balance the needs of both water users and the environment. In stressed catchment areas, there will be no new forestry developments, and government plans to levy water users (such as forestry) to finance future developments in order to alleviate shortages (Chalmers, 2001).

Environment

The 'White Paper on Sustainable Use of South Africa's Biological Diversity' (1997) issued by the Department of Environmental Affairs and Tourism (DEAT), emphasizes strict management of invasive alien plant species and makes specific reference to commercial forestry plantation species.

The 1998 National Environmental Management Act provides for a national environmental management system that will integrate with the national standards for sustainable forest management in the National Forests Act (Mayers et al, 2001).

Labour

The Congress of South African Trade Unions (COSATU) is an alliance partner with the African National Congress (ANC) government and has had significant political influence at macro and sector levels. COSATU-affiliated trade unions have engaged strongly with government in the design and implementation of its forestry privatization process.

The creation of a national forest estate

A brief history

1876 to 1910. South Africa's first plantation was established in 1876, in the former British-governed 'Cape Colony'. Plantations were an alternative to the small and fast-disappearing natural forests and expensive imported timber. With a growing economy, the Boer War and a rapidly expanding mining industry, the demand for timber was high. By 1910 the Cape Colony had 120,000ha of plantations, mainly state controlled.

1910 to 1950. Following unification of the South African republics in 1910, the new government began creating a national forest estate, initially driven by self-sufficiency and job creation. Saw-milling processing capacity also expanded, again primarily under state control.

From the 1940s, the government continued to expand the forest estate and establish processing facilities, but significant private-sector interest also emerged. Private processors built saw mills, and by 1950 plantations covered 693,000ha, of which 73 per cent was private by 1955.

1948 to 1972. The National party won power in 1948, and a 1956 Government Commission into Socio-Economic Development recommended that forestry be used as an economic development instrument in the new 'homelands'. During the 1960s and 1970s 150,000ha were planted in the Transkei, Ciskei, KwaZulu, Kangwane, Lebowa and Venda. Implementation became the responsibility of the homeland administrations, who used the process to create rural jobs, resulting in overstaffing and a lack of commercial focus, which is still the case.

From the 1950s to the 1970s, pulp and paper grew in importance. Private-sector afforestation focused on the pulp market. By 1972 the combined plantation area (including homelands) had reached 1.025 million hectares, of which 75 per cent was private.

1972 to 1994. An Afforestation Permit System (APS) was introduced in 1972 in response to growing concerns over the impact of uncontrolled afforestation on water resources. The APS restricted afforestation in stressed catchments and prohibited planting in sensitive locations, such as wetlands and close to water courses. These conditions remain essentially in force today; however, by 1994 the plantation estate had risen to 1.4 million hectares, mainly developed by the private sector for pulp and paper. The expansion of the pulp and paper sector, particularly for export, was encouraged through tax incentives and a General Export Incentive Scheme – which was removed after 1994.

Management of state-owned plantations

As a result of South Africa's unusual history, the privatization policy and strategy had to accommodate two very different state-owned plantation estates in the homelands and the former Republic of South Africa (RSA).

Homeland plantations

Different homelands administrations established and managed two main types of plantation:

1 Large softwood plantations of at least 300ha to supply saw mills, which were offered cheap timber to encourage investment in processing in rural areas.
2 Smaller hardwood pole plantations for building materials and firewood.

The quality of the plantations varied from excellent to very poor depending upon the sites, silvicultural and business management, and the level of conflict over land rights with local communities.

The entire homeland plantation resource (some 155,000ha) was transferred to the national Department of Water Affairs and Forestry (DWAF) following reunification in 1994.

Republic of South Africa pre-1994 (RSA) plantations (SAFCOL)

RSA-owned plantations pre-1994 were managed by the national Department of Forestry to supply industrial processing operations (predominantly softwoods for saw-milling and hardwood poles for the mining industry). The scale of planting was greater than in the homelands and management standards, site selection, and silvicultural standards were high.

In 1992 these operations were transferred to a specially created state-owned company, the South African Forestry Corporation (SAFCOL), as a prelude to privatization. In addition to managing about 262,000ha, SAFCOL owned and operated five saw mills and two telegraph-pole plants.

Use of state-owned plantations

Saw logs

SAFCOL and DWAF plantations produce about 4 million cubic metres of saw logs – 66 per cent of national supply. Saw-milling uses 4.4 million cubic metres of logs to produce 1.87 million cubic metres of timber. About 60 per cent of sawn timber is made into other products, 4 per cent is exported and the balance goes into structural timber and the construction industry. Pine thinnings are supplied to the pulp mills or board manufacturing plants. Saw mills range from log intakes of 200,000 cubic metres per year to less than 5000 cubic metres (Mayers et al, 2001).

Pulp and mining timber

Some state-owned plantations produce hardwood timber for the mining industry. This is increasingly diverted to pulp mills as mining operations find substitutes. There is also a pattern of conversion from pine to gum production, particularly in areas such as the KwaZulu Natal coast, where plantations are near the pulp mills and pulp production is much more profitable.

Poles

The pole market for treated and untreated poles for transmission and telephone poles, agricultural uses and construction consumes 750,000 cubic metres, split equally between gum and pine (LHA Management Consultants, 2001). About 25 per cent of the poles are from state-owned plantations.

Local construction

In some areas, particularly the Eastern Cape, communities depend heavily upon state plantations for building materials, mainly poles and lathes for traditional wattle and daub construction. Eucalyptus species and wattles (*Acacia mearnsii*) are preferred, the latter widely sourced from natural 'wattle jungles', often escapee plants from state forests or private farms. Eucalyptus poles are from large commercial plantations with areas of gum managed for community use, or from specially planted community woodlots.

Firewood

State plantations are an important source of firewood. Traditionally, communities have been allowed to collect head loads of deadwood and this right is enshrined in the New Forest Act. State plantations also support an important firewood industry. Traders collect firewood to sell to local markets and are charged per truckload.

Non-timber forest products

State plantations supply other non-timber products, too. On a small commercial scale there is resin-tapping on coastal *Pinus elliottii* stands, collecting *Casuarina* species needles for making turpentine substitutes, mushroom picking, and collecting ferns and flowering plants for florists.

At a household level, state plantations (and their adjoining indigenous forests) are an important supply of traditional medicinal plants, thatching grass, water and open areas for grazing.

Environmental and social benefits

State plantations and associated indigenous forests (where well managed) protect catchments, conserve soils and maintain stream flow and water quality. The entire SAFCOL estate is Forest Stewardship Council (FSC) certified and DWAF made good progress towards certification before privatization.

Evidence suggests that state plantations have helped to protect indigenous forests from overuse. Plantations often restrict access to indigenous forests and are an alternative source of products.

The role that all forests play in 'recreational, cultural and spiritual use' is also recognized in the Forest Act. The public has rights to any state forest for these purposes subject to reasonable management control to protect against fire and to secure health and safety requirements.

Securing local use rights

Neighbouring communities have enjoyed access and use rights to state forests for many years. Although there was a tariff system for certain products, it was unevenly applied.

The history of community access and use rights is not fully recorded, but it is recognized that these benefits should continue after privatization. They are protected as the plantations will still be state forests, and so will remain subject to the National Forest Act. The act provides for:

• free access for cultural spiritual and recreational use;
• licensing arrangements for any commercial use of a state forest (including commercial firewood collection or harvesting of any other product);
• an arrangement whereby local communities do not need a licence to collect forest produce as long as it is for domestic consumption.

Different perceptions of plantation forestry

Environmentalists
There is a small 'anti-forestry' environmental lobby in South Africa. Their concerns centre on two issues: firstly, the threat posed by plantations to bio-diversity conservation (particularly South Africa's grasslands) and, secondly, water supply and the threat to other land uses posed by forestry.

Rural communities
Rural communities' main interest in plantations is jobs. As a secondary concern, user groups have a direct interest in access to firewood building materials, medicinal plants and so on. The concept of rent or a share in profits from plantations is not familiar to rural communities.

The forest industry
The forest industry feels that plantations are an engine for economic growth and job creation in otherwise very poor rural areas with few opportunities. Forestry would be seen simply as a crop and as a raw material supply to wood fuel processing.

Public opinion
In South Africa, public opinion generally supports plantations from an economic, leisure or tourism point of view. Forests certainly form a backdrop to many tourism activities, such as hiking, fishing, camping and hunting.

Large international players

Sappi Forest Products owns and manages 490,000ha of plantations in Southern Africa. In 1998 they supplied 50 per cent of the fibre required to produce 1.8 million tonnes of pulp. Major acquisitions in the 1980s and 1990s made

Sappi a world player, manufacturing 5 million tonnes of paper and 3 million tonnes of pulp on three continents; 85 per cent of its sales and 70 per cent of its US$6 billion assets are outside South Africa.

Mondi is jointly owned by Anglo American, De Beers and AMIC, and manufactures pulp, paper and solid wood products. Mondi manages 440,000ha of plantation and exports 40 per cent of its production. Since the late 1980s Mondi has been acquiring shares in international companies to develop a global presence (Mayers et al, 2001).

Medium-sized players

There are an estimated 1800 (primarily white) private commercial growers who own around 364,000ha (24 per cent of the total area), making them collectively significant.

This group includes companies who grow timber to supply their own processing plants, including Masonite, which owns 18,000ha, feeding a hardboard mill, and Hans Marensky, which has recently taken over 60,000ha of state plantations to supply its saw mills in the Eastern Cape (Mayers et al, 2001).

Small or micro-growers

There are nearly 19,000 small or micro-growers in South Africa, holding woodlots averaging 2ha, totalling around 43,000ha in extent. More than 12,000 of these growers participate in company-sponsored out-grower schemes, including Sappi's Project Grow and Mondi's Kulanathi. Nearly all of these growers are in KwaZulu Natal Province near the pulp mills where there is a traditional land tenure system that suits the allocation of plots to individual households for tree growing (Mayers et al, 2001).

Regulatory and incentive framework for private-sector forestry

Regulatory environment

The 1998 National Forest Act seeks to create an enabling environment for the sustainable management and development of forests. It created a framework for the privatization of state forests and requires the minister to develop criteria, indicators and standards for sustainable management of all forests.

The most direct regulatory instrument governing forestry is the water-use licence issued under the 1998 National Water Act (see below). The private sector feels that the administration of this licensing process needs to be desperately improved. Forestry South Africa claims that the cost of applying and the processing time are major barriers to forestry development for both large and small growers. Ironically, the same department administers both water and forestry.

Incentives for forestry

There are currently no direct government incentives for forestry. General tax provisions allow companies to write off income from other activities against their forestry (or any agricultural) interests.

Forestry South Africa says that direct incentives are not wanted. In the past they distorted the market and reduced efficiency and competitiveness (Forestry South Africa, Mike Edwards, personal communication). Forestry South Africa sees the following priorities for government in the forest sector:

- Streamline the water licence system.
- Use the privatization process to improve efficiency and representation.
- Facilitate partnerships between companies and communities by addressing barriers, such as provincial politics, land tenure arrangements and water-use licences.
- Provide 'bio-security' by controlling the import and spread of tree diseases.
- Research the hydrological impacts of forestry compared to other land uses.
- Create a flexible and employment-friendly labour market and an investment-friendly tax regime.
- Invest in transport infrastructure.
- Raise the profile of forestry.

Changing demands on state-owned plantations

Government intends to privatize its plantations. Considering the pressures on South African forestry as a whole, how will these plantations be affected? The industry's success and direction will be fundamentally determined by its ability to respond to:

- international market trends;
- domestic economic factors;
- forest policy and legislation;
- policy changes in other areas (Mayers et al, 2001).

International market trends

Since the 1950s, South African forestry has been focusing on the international market. With increasing international exposure, South African companies are experiencing:

- price volatility;
- competition from countries with heavily subsidized industries (e.g. Canada) or whose resource is undervalued (e.g. the former Soviet Union);

- shifting trade patterns (e.g. the declining Japanese market and strengthening Chinese market);
- requirements for certification, particularly from Europe and North America;
- exchange rate fluctuations, (extremely beneficial to South African exporters for the last 18 months, but now reversing).

The net effect of these pressures (on both private and state forests) is to keep silvicultural standards high, operating costs low and to foster the development of certification to meet growing market demand.

Domestic economic factors

Investment in pulp plants

South Africa produces cheap pulp, most of which is exported. However, there is limited potential for increasing pulp production capacity in South Africa. The one area that could sustain enough increased afforestation to support increased pulp capacity – the Eastern Cape – is underdeveloped in terms of infrastructure and the potential forestry land is communally owned. There is uncertainty regarding whether communally owned land will be made available for forestry expansion in the future.

The saw-milling industry

By the early 1990s saw-milling was something of a 'poor relation' to the profitable, efficient and capital-intensive pulp and paper industry. This was largely because of past inefficiencies caused by the old government 'evergreen' timber-supply contracts. The 1998 Forest Act brought to an end the old timber supply contracts. Most have been renegotiated, while others are still being fought in the courts (Chalmers, 2001).

Log prices and labour comprise up to 80 per cent of solid wood processors' costs. As government log prices have risen towards international parity, the following efficiency improvements have occurred:

- *Improved recovery within saw-milling operations.* Companies are investing in better technology to extract maximum value from increasingly expensive logs.
- *Integration between plantation management and saw-milling* in both state plantations and the private sector. Integration cuts production and transaction costs.
- *Contracting out of forestry functions.* Contractors are now involved in transport, harvesting and routine silvicultural operations. Workers are paid about 60 per cent of the rates paid by the corporate sector and 25 per cent of state rates. Companies enjoy the reduced exposure to labour unions and no longer need to provide infrastructure and services, such as housing, schools, clinics and transport.

Forest policy and legislation

The privatization of state plantations will have an enormous impact on their future management and role in the forest sector. Privatization objectives require state plantations to deliver significant social as well as economic benefits. These objectives are discussed in more detail in 'Balancing acts: Reconciling public-policy objectives and private-sector investment' below, but include:

- improved efficiency and investment;
- broadening ownership to include the previously disadvantaged;
- recognition of underlying land rights;
- certification.

Policy changes in other areas

The two policy sectors expected to have the greatest impact on forestry are water and land:

- Water policy will impact upon production costs and the extent of future plantation expansion.
- The outcome of land reform will determine the long-term role of communities in the forest industry. If communities receive tangible benefits from forestry, it may encourage the release of more communal land for afforestation.

The approach to privatization

Early moves towards privatization: Commercialization and corporatization

Prior to the 1994 elections, a route to privatization had been envisaged and entailed three distinct steps:

1 *Commercialization* – adopting private-sector accounting practices and other working procedures while remaining in the public sector.
2 *Corporatization* – excising the commercialized operation and its workforce from the formal public service, and transferring it to a specially created, wholly state-owned company.
3 *Privatization* – selling shares in the corporation to the private sector.

This three-stage transition process encourages efficiencies without drastic or rapid change and allows the government to prepare the assets for sale in order to improve the sale price.

Commercialization began when the RSA's Forestry Department adopted a 'trading account' in 1985. Commercial accounting systems were introduced, along with budgeting practices that enabled the department to identify timber income and production costs separately. They were also allowed to retain revenue rather than returning it to the exchequer.

Corporatization began in 1989 with a draft bill creating a National Forestry Corporation. Following consultation with the forest industry, legislation was introduced in 1992 to corporatize the RSA's plantation assets. In September 1992, SAFCOL was incorporated as a public company and a board was appointed. In 1993, agreements were reached between SAFCOL and the government to transfer assets and staff (Mayers et al, 2001).

Post-1994 developments

In 1994 the former homelands' plantations became part of DWAF. In 1998 the government decided to privatize both SAFCOL and DWAF's plantations in a single phased process with three categories:

1   SAFCOL/Category A: the SAFCOL plantations (386,476ha) combined with approximately 70,000ha (50 per cent) of the DWAF former homeland plantations.
2   Category B: DWAF's remaining commercially viable plantations (approximately 70,000ha).
3   Category C: approximately 110 small scattered plantations (or woodlots), totalling only 17,000ha, established to provide communities with building material and fuelwood.

Initially SAFCOL–DWAF category A assets were divided into seven 'packages', each a logical business unit. Investors were invited to bid for a 75 per cent shareholding (of which at least 10 per cent needed to be black-owned) of each package. Special purpose vehicles (SPVs) were established for each package to facilitate the sale of shares. Minority shareholdings in each SPV are held by government (6 per cent), workers (9 per cent) and the National Empowerment Fund (10 per cent).

Government also determined that plantation land should remain in public ownership, and investors would be offered the use rights through long-term leases (Mayers et al, 2001).

Balancing acts: Reconciling public-policy objectives and private-sector investment

Objectives of privatization

The 1996 Forestry White Paper called for radical changes in the way that forests were viewed, valued and managed. Government would create the conditions necessary for others to manage forests in the national interest and

Table 10.1 *Status of South Africa's 'category A' assets offered to bidders during 2000–2002*

Package	Total lease area (hectares)	Total planted area (hectares)	Present status (2002)
Eastern Cape North	75,487	57,715	Sold to Singisi Forest Products Consortium, August 2001
Komatiland (Mpumalanga and North Province)	209,372	139,082	The package is to be re-tendered for the third time
Amatola (Eastern Cape South)	25,417	4,399	Rance appointed as a preferred bidder – negotiations at a final stage
Mountain to Ocean (Western Cape)	161,912	87,978	A 20-year exit strategy was established to withdraw forest production in the area; viable areas subject to tender process
KwaZulu Natal	43,946	32,652	Sold to Siyaqhubeka Consortium, September 2001
Total	516,134	321,826	

Notes:
1 Originally, seven packages were put out to tender. Komatiland was originally split into the Northern Province and Mpumalanga. No successful bidder was identified at the first attempt, so the two packages were combined and reoffered as Komatiland.
2 The Southern and Western Cape packages were also originally put out separately. After no bidders were found, it was decided to combine them as the Mountain to Ocean (MTO) package. Large areas (almost half the package) were identified as unsuitable for forestry in the long term, and this area will be clear-felled and converted to other land uses.

to regulate their actions. This withdrawal by the state mirrored a national programme of economic reform and liberalization in post-apartheid South Africa involving transport, manufacturing (including defence), telecoms, and public-sector infrastructure development and management (including roads and hospitals).

Several forestry-specific issues informed the privatization policy and are common to other privatization processes internationally:

• a belief that privatization will attract investment and expertise to revitalize assets that are perceived to be suffering from chronic under-investment;

- a fiscal imperative to reduce the burden on the state of subsidizing inefficient government plantations (SAFCOL made an operating profit within two years; but homeland plantations continue to operate at a significant loss);
- a recognition that managing plantations potentially conflicts with the government's regulatory role.

Privatization also offered a unique opportunity to achieve other White Paper policy objectives in a meaningful way, including the need to:

- Achieve a wider, more representative pattern of ownership in the forest sector, particularly amongst South Africa's historically excluded groups.
- Consolidate forest resources, often artificially and inefficiently split.
- Improve efficiency in processing industries by increasing effective competition for raw material supplies.
- Formally recognize the land, access and use rights of rural communities, many of whom were dispossessed of these rights when these plantations were established.
- Facilitate black empowerment through direct participation in forestry operations, training and skills development, affirmative action in management, and entrepreneurial opportunities through outsourcing, partnerships, procurement and easier access to financing (Mayers et al, 2001).

Managing trade-offs within the privatization process

Government has to prioritize amongst mutually exclusive policy objectives; for example, attracting foreign investment may conflict with encouraging local black economic empowerment and the trade union movement will emphasize job security, while the Treasury emphasizes financial return on investment. These trade-offs are political. The solution needs to be politically acceptable, economically sustainable, legally implementable and 'value for money' for the state. In South Africa, some bids won not on price but because they accommodated public-policy objectives.

Compliance with the various terms and conditions of the bid can be located and enforced in:

- the transaction documents, such as sale of business agreement, shareholder's agreement and sale of shares agreement; these deal with the business-specific issues;
- the lease agreement that governs the use of the land;
- legislation including the Forest Act, Companies Act and land tenure legislation that provides a regulatory framework which does not need to be duplicated into transaction documents.

Bid evaluation

The government need not select the bid with the highest price. In South Africa bidders submitted bids against a set of qualitative and quantitative criteria, striking a balance between public- and private-sector objectives.

Specific trade-offs and how they were managed

Secure tenure rights for bidders versus protecting underlying land ownership and informal tenure rights

One of the most challenging aspects of the privatization process was securing reasonable tenure security for the bidder to encourage investment, while at the same time protecting communities' underlying land and other informal tenure rights. The first decision was to lease the land rather than to sell it.

The New Zealand Crown Forest Licence was a template for the original lease drafts, as it also combined a land claims process with privatization. The draft lease includes the following:

- The tenant receives a lease with an effective minimum 70-year duration.
- The tenant pays a market-related rent to use the land. The value of the standing tree crop is not taken into account in determining land values. This rent is held by DWAF in a trust fund until the underlying landowner is identified. At that point, any accumulated and all ongoing rent is transferred directly to the landowner.
- The tenant receives full undisturbed possession of the land subject to the requirements of the Forest Act that allows public access for cultural, spiritual and recreational purposes.
- The tenant is issued with a licence to operate.

The leaseholder's blanket licence covers all activities, including silvicultural operations, ecotourism, quarrying, hunting and so on. The leaseholder may then issue licences to third parties for some of these activities. These key lease features are non-negotiable. In the event of a successful land claim after the lease has been signed, the claimant receives title to the land and the rent, but on condition that the lease remains in place, which provides adequate security to the private-sector operator. The claimant can challenge this in the Land Claims Court. If they win, then the state must compensate the leaseholder for any losses incurred.

Short-term job losses versus long-term economic sustainability and growth

Unemployment is reaching 40 per cent in South Africa. It is a critical issue. Short-term job losses are likely to be more important to many than promises of long-term sustainability and economic growth. In order to mitigate negative social impacts, the following mechanisms were employed:

- All SAFCOL workers are being transferred to the new employer on existing terms and conditions of employment.
- An industry norm number of DWAF workers will transfer at SAFCOL wages and terms and conditions. DWAF workers who are in the process of transferring will receive a transfer package to compensate for the reduced salary.
- Employee Share Ownership Plans (ESOPs) are promoted to ensure that employees have a real stake in the plantations.
- There will be a moratorium on retrenchment of transferred workers by the new company for three years.
- Bidders' human resource development plans had to address affirmative action and effective human resource development planning.

When workers leave (voluntarily or otherwise), they are eligible for a social plan. This is a package of training and counselling services to help workers deal financially and psychologically with unemployment and help them to retrain for re-employment or self-employment. Most workers, to date, have chosen to learn the skills suitable for self-employment at a micro or household level.

While it is too early to judge the Social Plan Framework, there is still much to do to manage the short-term negative impact of restructuring. Additional long-term support is needed to assist these workers to maintain their activities sustainably into the future.

Black empowerment versus established white-owned enterprise (small versus large)

The category A process was designed to attract large investment. A few smaller investors formed consortia amongst themselves or with larger companies.

For category A and SAFCOL transactions, a minimum of 10 per cent of the shares of the bidding company had to be black-owned. Singisi Forest Products created a community trust through consultation with the community as their black empowerment partner. The preferred bidder for some transactions was a 100 per cent black-owned consortium, including a number of small forestry-related companies and entrepreneurs.

In addition, 10 per cent of shares were transferred by government to the National Empowerment Fund (NEF) to facilitate access to state assets by the previously disadvantaged. The NEF Trust receives shares of all state-owned enterprises undergoing restructuring. It will promote empowerment by marketing investment units to historically disadvantaged individuals.

Government is keen to support further community trusts in all other plantations. Category C transactions (community woodlots) will likely be taken over by community trust structures. DWAF will support business planning for these plantations and a Forestry Enterprise Development Office is being established to facilitate further afforestation on communal land and to help communities own and manage forestry-related enterprises.

Enforcing environmental management standards

The lease is an incentive for sound management. Not only does the tenant have rights almost as strong as freehold, but the lease is also a mortgageable asset. Regulation by government, however, is still required. The system for monitoring and enforcing management performance against standards needs to be effective, but also practical and cost effective.

The solution was to require the leaseholder to obtain certification from a government-approved body within two years of obtaining their lease – a neat, cheap way to monitor management against sustainability criteria.

Best-bet practical instruments and processes

Create a robust policy and legal framework for privatization

Once policy is in place, an enabling legal framework for implementation can be created. The new Forest Act created an opportunity to start over completely rather than amend existing legislation.

Policies and legislative frameworks of other relevant government departments, such as land reform, need to be taken into account to develop a complementary process.

Create the capacity to manage the process effectively

Many privatizations fail to achieve their objectives, not because they were wrongly conceived, but because they were poorly executed. The South African experience has depended upon the following:

- access to technical expertise (local or bought in) in a number of core areas, such as forestry management, wood processing, finance, and legal and project management;
- process management expertise, whether to manage the process itself or to manage the bought-in experts against a timetable and a budget;
- clear governance arrangements; the challenge for DWAF and the Department of Public Enterprises (DPE) is to maintain objectivity and weigh up what is in the state's interest, compared to what is in the interests of any one party;
- a mature, transparent and accountable civil service culture;
- a dedicated budget against a reasonably flexible timetable. Strong political support

Civil servants can administer a process of this nature up to a point. However, political interventions will be needed in order to break deadlocks around key issues, so both political champions and politically astute civil servants are required to bring such a process to a conclusion.

Provide adequate time and space for consultation with and 'buy-in' by other relevant stakeholders

Groundwork must be timely and thorough in order to accommodate all relevant stakeholders, secure buy-in and identify possible risks up front. The principles that underpin such an interaction should be embraced by all involved. The following issues are key to a successful process:

- Consultation with other government departments needs to take place at several levels and over a long period of time.
- Open consultation with bidders about government expectations is important. The process should allow the private sector to come up with creative ideas on how partnerships can be established.
- Trade unions should be involved from the beginning. There may be an existing culture of engagement and bargaining forums already in place.
- The national Treasury needs to be kept constantly informed about the process from beginning to end.

Institutional capacity-building support to facilitate participation by previously disadvantaged groups

Communities and small enterprises need to be well organized and have access to independent advice, such as lawyers and transaction advisers, or business management training, if they are to participate effectively in partnerships and negotiate favourable terms for themselves.

In the long term, institutional support can be provided by the well-established private-sector partner within the consortium. Sound institutional arrangements and provision for institutional development and support are key to ensuring that they play a developmental role and that trustees remain accountable to their members on the management and distribution of benefits.

Create the capacity to manage government's post-transaction responsibilities

The privatization process leaves behind a number of residual regulatory responsibilities for government. DWAF has created a dedicated land management section to manage the leases in terms of financial management, performance monitoring, and dealing with such issues as rent review or applications for a change of use. This unit sits within the directorate: regulation which has the overarching responsibility of overseeing the implementation of the new Forest Act.

The Department of Public Enterprises oversees other transaction commitments and is responsible for the long-term share allocation arrangements, such as the ESOPs and NEF. They are also responsible for monitoring compliance with bidders' investment commitments.

Conclusions and ways forward

Key lessons

The key lessons learned are as follows:

- Set clear, politically acceptable objectives through consultation with key stakeholders and reference to sector and macro policy.
- Communicate the objectives clearly and early to potential bidders and take feedback on their reaction to determine the feasibility of achieving an acceptable transaction.
- Maintain dialogue with key stakeholders as the process unfolds and as delays inevitably arise.
- Use the various instruments within the transaction to achieve the multiple objectives:
  - Use the bid process as a market instrument and evaluate bids (transparently) to select a preferred bidder.
  - Use the sale of business agreements to secure commitments to economic development, including downstream processing.
  - Use the lease as the central instrument to transfer to the private sector use rights over state forest land (and avoid loading the lease with other issues that can be covered elsewhere).
  - Use existing legislation to regulate. Do not over-regulate by encumbering the lease, and try to allow the leaseholders to operate on a level playing field with other private companies, all of whom must operate within the law.
- Provide incentives for sustainability rather than regulate for it. If you use the lease to provide the private sector with long-term security to yield a return on its investment and the right to trade its investment, then it has an incentive to manage the resource to its full potential, rather than deplete it.
- A requirement for certification in the lease enables government to transfer much of the cost of monitoring and reporting on sustainability to the private-sector operator.
- Create adequate capacity to manage government's residual responsibilities in terms of the lease and other transaction commitments.

Impact of privatization: Gains and losses

The true success of any privatization can only be measured many years after the process has been implemented. South Africa has only just started. However, it is felt by those close to the process that the transfer of management responsibility from the state to the private sector has been conducted in an inclusive and appropriate manner and that the transactions that have arisen from the exercise provide a robust platform for long-term sustainability.

Expected gains

Added impetus to industry development. There appears to be industry enthusiasm about privatization (although this may have been dampened from time to time by various delays to the process).

Reduction in public debt. Government (at least in terms of the homeland forests) is very likely to have to unburden itself from the financial liability of loss-making operations, leaving more money for more essential services.

Employment. In the short term, jobs will be lost, salaries will drop, and terms and conditions will be less favourable; but those jobs were not sustainable and the new levels should be if plantations are well managed.

Increased productivity and efficiency. The homeland forests could easily increase productivity by 50 per cent, which will feed increased processing capacity and provide more jobs.

Further afforestation. Privatization could be a key to unlocking communal land for commercial afforestation in the Eastern Cape. This would have significant benefits in terms of jobs and income to communities.

Social capital. It is argued that even though the rental flows from forestry will be small (relative to the needs of communities), the institutional capacity created to manage these income streams will provide a useful platform upon which other local economic development initiatives can be managed.

Recognition of land rights. By recognizing land rights and channelling income to communities, the process is addressing, to some extent, some of the injustices of the past. In future, communities may negotiate their own deals with forestry companies.

Black empowerment. If black people own forests and the companies that manage them, then once again a contribution will have been made to addressing the injustices of the past.

Losses

The short-term job losses will be the most obvious loss and will hit communities hardest.

There will certainly be some *reduction of forest area* in the short to medium term as government pulls out of non-viable operations, and meets FSC restrictions for planting riparian zones, etc. There will be some reallocation of timber supply away from timber processing companies who have not participated in successful bidding consortia. There will be some further rationalization of the saw-milling industry, as a result, and some companies will lose out.

If the *quality of management* deteriorates under the private sector, then privatization will be judged a failure. The world market for forestry is not strong at present, and South Africa needs to attract high-quality companies.

The work remaining

The privatization experience in this study relates to the disposal of SAFCOL and DWAF's category A plantations. The issues facing the B and C plantations are as follows.

Category B plantations include 14 packages of about 1300ha each – an opportunity for smaller investors. Some are borderline viable, but many are not and in some cases small businesses already depend upon them. Various approaches are being explored, including partnerships between provincial and local governments, communities, workers, small emerging forestry businesses and large established companies.

Industry feels that it is unreasonable to expect category B bidders to achieve certification (based on FSC standards) within two years. An innovative transitional arrangement is needed to allow smaller, less well-resourced companies to bring management standards up to an acceptable level from their current very low base.

Category C plantations were established to supply communities with building materials and fuelwood, and government seeks only to transfer them to communities, along with training to help communities to manage and develop their own forestry resources.

Residual responsibilities

In addition to dealing with the last of the state-owned plantation, the privatization process will continue to develop and maintain the capacity of the private sector to manage the resources sustainably, and the capacity of the state to monitor, regulate and, where necessary, provide support.

11

From Plantation Developer to Steward of the Nation's Forests: The UK

David Grundy

Introduction

The UK is a densely populated industrial country with a large demand for wood products, but only small domestic wood production. There is a small, modern forest-based industry producing a wide range of paper products and wood products. Wood imports meet over 80 per cent of consumption. The inescapable high dependence on imports has, in the past, raised economic and security issues. As detailed below, these are now less pressing.

The contribution of forestry and the industry based on home-grown wood to gross domestic product (GDP) is relatively small – indeed, too small to warrant separate identification in the national accounts. Total employment derived from UK forests at some 30,000 is not significant in a national context (Forestry Commission, 2001). But forestry employment is important in some remote, sparsely populated areas of Scotland and Wales, which have not benefited from recent changes in the demographic trends in rural areas, where in-migration of businesses and people is beginning to offset the effects of depopulation and relative poverty. For most rural areas, forestry is not now a significant source of employment, and in these its potential is too small for it to be a target of policy.

Over 90 per cent of the land is owned privately. There are no legal obstacles to the sale of land; but the competition between uses – for agriculture, industry, communications, housing, defence and conservation – is a key factor in deciding the scale and location of forestry. In particular, domestic demand for food products, supported by high prices engineered by the European Union's (EU's) Common Agriculture Policy (CAP), has favoured agriculture.

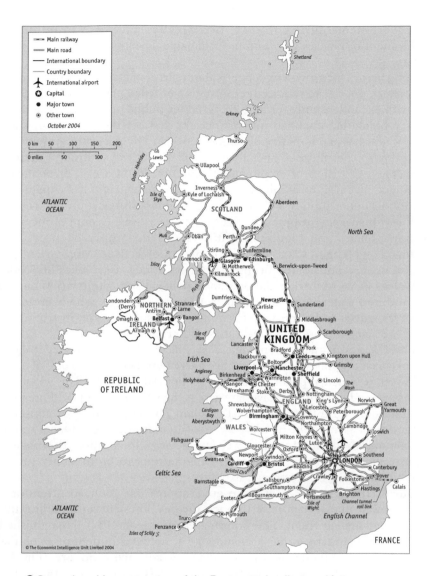

Source: © Reproduced by permission of the Economist Intelligence Unit

Figure 11.1 *The UK*

In contrast, forestry offers relatively low commercial returns over very long timescales. Left to itself, the market would invest little in wood production. Forests, woods and plantations cover only 11 per cent of the land area, nearly all of which was originally forested.

There are well-developed environmental policies backed by legislation to protect wildlife, landscape and water. These restrict the scope for new plantations, and increase the costs of exploiting the existing wood resource.

UK governments during the last two decades have also promoted the development of policies to promote sustainable development, which apply to forestry.

Long before the growth of concern about the sustainability of resource use, UK forestry policy had required reforestation of felled plantations as a general rule, with exceptions only for land required for public purposes. Nonetheless, forestry policy has changed much since the start of the last century when forestry first became a matter of public concern. The government now looks to forestry to contribute to social welfare in several ways: through increased biodiversity; provision of employment in remote rural areas; diversification of farmers' incomes; underpinning forest-based industries; and expansion of recreational opportunities. The state now promotes afforestation, reforestation and forest management practices that serve these aims. A suite of incentives, taxation policies and regulations, discussed below, have been gradually developed and refined to target this complex of objectives. Most recently, the devolution of responsibility for internal affairs to new assemblies in Scotland and Wales has led to further elaboration of policy. Each country (i.e. England, Scotland and Wales) now has its own forestry strategy, which targets the objectives judged to be most useful in its circumstances. The instruments of policy are being modified to reflect these differing strategies.

The UK government, nonetheless, is still involved in forestry policy. It controls taxation policy, which can have a large impact on the value of a long-lived asset. It is also sponsor of the UK Biodiversity Action Plan, which aims to conserve and enhance biodiversity across the whole of the UK. Forests and plantations are an integral part of this plan. The importance of conserving the biodiversity inherent in the very small remaining areas of natural and semi-natural woods and forests is well appreciated. In addition, there is much scope to increase the contribution of plantations to biodiversity through diversification of species and age structure.

During the last five years, the government has focused attention on achieving well-managed forests through the development and monitoring of standards of management. The Forestry Commission (FC) has published a UK Forestry Standard. It has also supported the development of a voluntary and wider set of standards, the Woodland Assurance Standard. The state and some privately owned plantations and forests are managed to meet this standard.

The gradual development of a complex forestry policy, with multiple objectives, can be seen as an attempt to satisfy a multiplicity of interests. Private owners of plantations have sought economic benefits: income and capital appreciation. They have also used their plantations for field sports. The forest-based industry has been most concerned to secure a reliable supply of suitable wood. Many people have taken up the opportunities plantations offer for outdoor recreation. In the densely settled south, where most private land is closed to the public and there are only rights of way available for public access, the state-owned woodland and plantations have been much used for the simple pastime of walking the dog. Farmers have been reluctant to see land taken out of agriculture, fearing a weakening of the local farming economy; but they took advantage of the opportunity to create small woods on their farms when the government introduced a scheme tailored to their circumstances. Similarly, environmental non-governmental organizations (NGOs) have opposed the conversion of upland pasture to plantations of non-native species of trees, but have welcomed planting of new native woods and forests and encouraged owners to develop the potential of the plantations for biodiversity. The history of state-owned and private plantations reflects the interactions of these interests.

A brief history of state-owned and private plantations

The centuries-long and almost uninterrupted trend of deforestation in favour of agriculture, industry and settlements had led, by 1900, to only 5 per cent of the UK land area being under trees, with an industrial economy almost completely dependent upon wood imports. World War I exposed the risks of relying upon imports for what was then an essential industrial resource. The main motive for establishing, in 1919, a state agency, the Forestry Commission, to promote afforestation was thus national defence. A secondary reason was the opportunity to increase rural industry and employment at a time of economic malaise.

From the outset, the FC set ambitious targets for reforestation and afforestation. To promote restoration of depleted forests, the FC provided advice and encouragement to private owners who could qualify for management grants if they undertook to manage their forests in an approved manner. The FC was given power to control felling through a licence system to prevent further loss of standing trees.

The FC took the lead in developing the techniques of plantation establishment, management and exploitation. It found that North American conifer species grew well in the moist British climate. Only marginal land of very low productivity was available on a large scale for afforestation; the FC developed methods of getting trees to grow despite nutrient deficiencies. At the beginning, the practice was for foresters to make a careful choice of species to match the site, using native species of herbs and grasses as indicators of soil quality. Later, however, it became possible to change the soil's physical and chemical properties – through draining, cultivation and fertilizing – to grow the most productive species, usually Sitka spruce, Douglas fir, hybrid larch or Corsican pine. This radical simplification of forestry practice increased yields, reduced net costs and improved the marketability of the wood produced. But, later, it also led to conflicts with a public becoming increasingly concerned about environmental damage to the countryside.

The FC aimed to meet the targets for afforestation by buying land for planting itself, and through incentives for private owners to plant. These planting grants were needed to compensate for the low rate of return on afforestation. As there was negligible planting without grants, the FC was able to control the nature and location of afforestation, which was not subject to the planning controls on development exercised by local government. However, for much of the period, planting on productive land was prohibited, reflecting the priority given to policies for increasing domestic food production.

Thus, the FC was able to control felling and planting and influence management. Over time, in response to increasing concern about the scale of land-use change and its impact on the countryside, with growing opposition to large-scale afforestation with non-native conifers, a set of controls was elaborated to mitigate environmental damage. The FC developed a system of consultation with local governments, agriculture departments, environmental agencies and NGOs, and, eventually, with the public generally, in an attempt to reconcile conflicts. Mandatory environmental impact assessment of large-scale afforestation proposals was introduced in 1989.

These policies, to a degree successful, came under increasing strain during the 1980s because the rate of private-sector afforestation increased rapidly, the result of a very favourable tax regime, under which the costs of afforestation were very largely subsidized by the Exchequer (i.e. the taxpayer). Like the FC at that time, the private sector concentrated mainly on large-scale afforestation with non-native conifers. With tax relief as the main driver, the official incentive scheme offering planting grants was reduced to a minor supplement; its main function became the control of an engine that seemed to those concerned about the conversion of unimproved land to have become an unstoppable juggernaut. Forestry became at this time a focus of bitter conflict.

The government reviewed the tax and grant regimes for private forestry in 1988. It eliminated the powerful tax incentive for planting (under which forestry operations qualified for tax relief while income and capital appreciation were almost tax free) by the simple expedient of taking forestry out of the income tax system. To compensate for the large reduction this produced in the incentive to plant, the planting grants were increased substantially. These were tailored to promote a much wider range of plantation types: there was more emphasis on broadleaves, small-scale woods, community woods offering public access, and recreation of woods and forests of native species. Correspondingly, there was less emphasis less on large-scale afforestation of unimproved land with non-native species.

The FC's own afforestation programme was sustained from 1920 until about 1990. The government financed the purchase of land and the costs of planting. During the years up to about 1960, the need to increase self-sufficiency against the possibility of a future war remained the main objective. It was then recognized that the great improvement in the relations between the large states of Western Europe, and the development of nuclear weapons, made this security objective obsolete.

However, afforestation continued. Forestry seemed able to contribute to employment in rural areas, increasingly suffering from depopulation and relative poverty as the agricultural workforce fell rapidly in response to the technological advances in farming. The same forces – very large increases in labour productivity resulting from mechanization (chainsaws, tractors, harvesters, etc.) – have now reduced forestry employment in a similar way.

Afforestation by the state came to an end with the advent of a government convinced that the private sector was a better manager of commercial businesses, which ought to be moved out of the public sector. The FC began a programme of disinvestment, discussed below.

The policy of a state-led partnership with the private sector succeeded in establishing over 1.5 million hectares of new plantations, mainly softwoods of non-native species (see Table 11.1). When added to the long-established woods and forests in private ownership, these have raised the area under trees to 11 per cent of the total land area. The private sector has about two-thirds of this. However, the main change has been in state-owned plantations, which increased from nothing to 0.9 million hectares – the biggest single change in land ownership and use in the UK in modern times.

Table 11.1 *State and private new planting in the UK since 1920/1921*

Five-year Period	State	Private	Total
1920/1921–1924/1925	15.3	12.7	28.0
1925/1926–1929/1930	32.4	12.7	45.1
1930/1931–1934/1935	34.3	10.8	45.2
1935/1936–1939/1940	39.1	10.0	49.1
1940/1941–1944/1945	20.3	7.8	28.1
1945/1946–1949/1950	46.7	19.5	66.3
1950/1951–1954/1955	82.7	29.2	111.9
1955/1956–1959/1960	71.7	57.8	129.5
1960/1961–1964/1965	85.4	63.2	148.6
1965/1966–1969/1970	79.8	62.4	142.3
1970/1971–1974/1975	102.6	95.8	198.3
1975/1976–1979/1980	74.5	39.4	113.9
1980/1981–1984/1985	45.0	66.1	111.1
1985/1986–1989/1990	22.8	102.5	125.3
1990/1991–1994/1995	11.1	78.9	90.0
1995/1996 –1999/2000	1.2	79.3	80.5

Notes:
1 Up to and including 1924, state figures include Crown woods under the technical supervision of the Forestry Commission.
2 Up to and including 1937, private figures include planting work done in connection with Proceeds Sharing Schemes initiated by development commissioners.
3 Up to 1967, the financial year ended 30 September. From 1969, the year was to 31 March; so the period of 1966 to 1970 above only covers 4½ years, from 1 October 1965 to 31 March 1970.
4 Up to and including 1970, private woodland figures are for all planting (new planting and restocking).

Wood production has increased from a negligible amount to 10 million cubic metres and now supplies 15 per cent of the domestic demand (Forestry Commission, 2001). The steady growth in wood production from the maturing plantations presented an opportunity to promote regional industrial development. Until about 1980, UK governments promoted industrial growth in economically depressed regions by offering capital grants to inward investment. Using these, and drawing on its ability to offer secure long-term supplies of wood, the FC was able to stimulate the development of a modern, large-scale processing industry. In this way, wood-based industries were attracted to exploit the new resource and its proximity to a large market. There are now 4 large pulp and paper mills, over 300 saw mills, and 9 wood-based panel factories, in addition to small-scale producers of fencing and other wood products.

As outlined above, the state came to own, through the FC, nearly 1 million hectares of mainly softwood plantations. It had been envisaged from the outset that the investment by the state in its own plantations would – eventually – produce a return. By 1990, the FC's wood production had become substantial and the domestic market was well developed. It seemed reasonable to the Treasury and the FC to look for a positive return. This was easily said; but it was difficult even to measure the return on investments made over many decades. There was also the complication that the FC carried out departmental, regulatory and promotional functions, as well as managing the state-owned plantations, with the staff dividing their time between these activities. Allocating costs between the commercial operations of plantation management and the non-profit making governmental functions was an arbitrary exercise.

A solution to these issues was found as an outcome of the review of the future of the state-owned plantations, discussed below. A new agency, the Forest Enterprise (FE), was carved out of the FC, and was made answerable to it for the management of its plantations with no distracting departmental, regulatory or other functions. Its function is to manage the state-owned plantations and forests to a set of objectives specified by the government for production, recreation, biodiversity and landscape conservation, as well as for efficiency and profitability. It contracts out planting, harvesting and other forest operations, employs contractors on the work it manages itself, and negotiates contracts with wood-using companies. It has withdrawn completely from manufacturing operations (which were never substantial) and it has divested itself of small and remote plantations that were relatively costly to manage. In general, it uses the private sector whenever the relative costs indicate that it would be beneficial.

The private sector has also increased its supply of wood to industry. It owns nearly 2 million hectares. Some are long-established broadleaved woodland of low productivity; but there is also a large area of broadleaved and conifer plantations that supplies wood to industry (Forestry Commission, 2001). Many of the plantations are owned as investments and managed by professional managers who use contractors for forest operations. As well

Box 11.1 Development of the Forestry Commission of Great Britain through the 20th century

1920s–1950s. The Forestry Commission was established. Its primary objective was to create a strategic reserve of timber that could be drawn from in the event of war. With powers to acquire and develop land, and to support private-sector afforestation with grants, the Forestry Commission estate comprised 1 million hectares by 1960.

1940s–1960s. Post-World War II concerns for protection of wild spaces, creation of national parks and increasing demands for outdoor recreation. The Forestry Commission responded with a recreation and social agenda (e.g. forest parks). The Forestry Commission estate comprised 1.19 million hectares by 1970.

1960s–1970s. Increasing provision for access and recreation. The impacts of forestry on the visual landscape began to be recognized. Landscape appraisal and forest visual design became standard practice on the Forestry Commission's estate by 1980. Private-sector afforestation took off, driven by tax incentives and grant aid. A major forest policy review concluded that government support for forestry was justified only after non-timber benefits were brought into the equation. The Forestry Commission estate comprised 1.26 million hectares by 1980.

1980s. Privatization agenda. The Forestry Commission began disposing of forests and reduced its afforestation programme; the private sector became the main afforestation agent. The environmental agenda began to develop as opposition to continued loss of semi-natural woodland and afforestation of ecologically valuable habitats became better organized. The Forestry Commission published the first in a set of best-practice guidelines, but was unable to enforce implementation in the private sector. Tax breaks for forestry were abolished; support to the private sector continued through grant aid conditional on compliance with best practice. The concept of forest management for multiple benefits became firmly established. By 1990, the Forestry Commission estate comprised 1.14 million hectares.

1990s. Forests' role as an agent for sustainable development began to emerge. Conflicts between the Forestry Commission's support and enforcement functions and its role as manager of the nation's public forests were addressed by establishing the Forest Enterprise as a separate division. The move to privatization was blocked by concerted opposition from non-governmental organizations; the Forest Enterprise was established as an agency of the Forestry Commission. Pressure on timber prices forced business efficiency, downsizing and a search for partnerships. Devolution to Scotland and Wales accelerated devolution within Forest Enterprise; programmes were developed at regional level in response to the agendas of the devolved administrations.

as encouraging the growth of demand for wood produced in the UK, the existence of a substantial state-owned plantation sector helped initially to develop private contracting businesses and provided experienced plantation managers. These services are now well established and no longer depend upon the state.

The main current influence of the FC on the private sector is through felling and planting controls, and grants for planting, restocking and management, which enable it to promote and regulate planting, felling and forest management. Much of the current new planting is aimed at biodiversity, landscape, sporting and recreational objectives. The FC also undertakes the bulk of the research into new technologies, which remains largely financed by the government.

Evolving demands

The long-run trends in the UK economy are for increasing predominance of services at the expense of agriculture and manufacturing. The UK's economy has come to depend less upon industry, and industry has become more economical in the use of raw materials, including wood.

Demand for wood in the UK grew slowly through the period of industrialization, and has now stabilized at 45 million to 50 million cubic metres a year, or about 1 cubic metre per head (Forestry Commission, 2001). UK wood production is increasing with the progressive maturing of the new plantations and now meets 15 per cent of total consumption. There is the potential to increase domestic production further, displacing imported wood products. However, prices are set by imports in a free market; recent trends are markedly down, with serious consequences for the profitability of commercial forestry.

The upshot is that forestry's importance as a supplier of raw materials has declined. Indeed, forestry is now so unprofitable as to raise issues about the value of plantations for solely commercial purposes. Tight control of costs is more important than ever.

In contrast, there has been a growing demand for outdoor recreation. With the growth of car ownership, leisure time and personal incomes, forests have become more valuable as recreational resources. The main demand is just for access for walking: in densely settled areas, private land is usually closed to the public. In addition, a wide range of traditional and new outdoor activities take advantage of forests – orienteering, car rallying, cross-country cycling, fishing, wildlife observation, skiing, horse-riding and trekking. Recent surveys show some 300 million to 400 million visits a year to forests in the UK (Forestry Commission, 2001).

The second major change in demand on forests has come from the huge increase in the perception of the value of environmental services, such as biodiversity protection, landscape beauty and carbon sequestration. The impacts have been complex, affecting afforestation and forest management.

Thus, land that is agriculturally poor, and was once seen as eminently suitable for commercial re-afforestation, is now appreciated for its intrinsic conservation value compared to the intensively farmed 'deserts' of modern agri-business; it is, at least, 'semi-natural'. It would be judged as more valuable ecologically in its current use as rough pasture than under a plantation of one or two non-native species of conifer. However, much of the area of rough grazing was once forest. Planting of trees to restore these native woods is generally welcomed. The environmental pressure is for recreating some of the lost indigenous woods and forests on farmed land.

The same pressure bears on plantations on ancient woodland sites. Where some of the original ecosystem remains, there are opportunities to replace exotic tree species with native trees, along with the reintroduction of the characteristic flora. The reduced commercial value of the timber has strengthened the case for restoration: there is less to lose.

The uniformity of the new plantations was much criticized. Lacking diversity, they were poor habitats for wildlife. They were also often intrusive in the upland landscape. Action taken to diversify tree species and age structure as the plantations are managed and regenerated, and to take opportunities to harmonize them with the landscape, has been welcomed.

The new plantations were found to have adverse impacts on water quality and quantity on some sites. They acidified rivers and streams by capturing acid particulates from the atmosphere and channelling them into watercourses. Unplanted buffer zones between the plantation and the streams running through it ameliorate this effect.

The improved drainage required for tree establishment has led to more rapid run-off with an increased load of soil, causing increased turbidity of watercourses. Again, techniques have been devised to slow down run-off and reduce this damage.

Tree planting of all kinds, including plantations of exotic species, adds to carbon sequestration. Under the Kyoto Protocol, additional woodland planted since 1990 contributes to the UK's carbon dioxide emissions target by removing carbon from the air. The amounts are relatively modest. In a developed and densely settled country, it seems that carbon lock-up by forests can have only a minor role, albeit positive and welcome.

More vaguely, but perhaps of most importance, there has been growing public appreciation of woods and forests as desirable in themselves. Charitable trusts aimed at improving woods and creating new ones have attracted voluntary support. Wildlife NGOs have become involved in forest management. Ownership of woods and forests by organizations and individuals for non-commercial purposes is increasing. The public is generally favourable; but opposition to large, uniform plantations of conifers has remained as a legacy of past forestry practices.

Along with the changes in forestry practice, there has been an enormous increase in public involvement in state forestry. FC managers consult panels of experts and other interests, many forests have some kind of partnership involving local people, and the FC itself is a partner with environmental

NGOs in a variety of projects. The upshot is that the FC is now trusted with the long-term stewardship of the land it manages; and this is an important factor in any debate about privatization.

The evolving demands are mainly for services provided by plantations outside the market. Their increasing importance contrasts with the reduced profitability of investments in plantations. Indeed, low profitability is a systematic problem for afforestation in environments where tree crops take 40 or more years to mature.

If the domestic market is open to trade with no or low customs duties (as it should be according to free-market economics), prices are set by imports. These will come from exploitation of natural forests and from environments where trees grow more quickly. In the former case, no initial investment in afforestation is required; and in the latter, it produces an earlier return. So, despite its suitability for forestry, the UK does not have a comparative advantage for afforestation as a commercial investment at present. Rates of return are low.

It follows that a subsidy is needed to stimulate planting by the private sector, and public finance must be provided if afforestation is carried out by the state. Both these need some justification. The history of British forestry shows that a range of 'public good' reasons has been called upon to support the case for state aid to increase the country's wood resource.

The original raison d'être for afforestation – national security – is no longer relevant. Forestry as a provider of employment is not now a main factor. Its importance as a source of raw material has declined, and so has the financial return from timber sales. But plantations are valuable as recreational resources. And they have the potential to increase biodiversity. Environmental pressures have increased the costs of operation and reduced the scope for new plantations aimed at commercial objectives. In contrast, they have favoured the creation and restoration of native woods (of low timber productivity), and modification of plantations to make them more diverse and, thus, more valuable as recreational and wildlife resources. But, again, this also increases costs and reduces productivity. Carbon sequestration is valued by the government, but does not contribute to the forest owner's income, although it may be considered to be part of the rationale for planting grants. The public attitude is generally favourable to forestry; but opposition to large-scale land-use change remains.

The broader context of economic policy has also changed. UK governments now follow free-market economic policies: operation of commercial businesses should be in the private sector; the state's role is to maintain a stable currency and to provide for regulation where needed. The regulator must be impartial and independent of the businesses that it oversees. Any intervention by the state to promote an activity should be justified by the public benefits it would achieve, benefits not being delivered by private agents acting in pursuit of their own interests. This emphasis on market economics has also influenced the roles of the FC and private owners.

Changing roles

For most of the 20th century, the Forestry Commission, as the state's agency for promoting forestry in a deforested country, was accepted as at first pioneering the development of forestry and then leading the forest industry. It developed techniques across the range of forest operations – nurseries, afforestation, forest management, harvesting and marketing practices – and sought to transfer them to the private sector. As the largest owner of commercial plantations, it took the lead in promoting investment in forest-based industries, mainly through contracts guaranteeing supply at competitive prices. The private sector generally welcomed the state's role as leader and provider of technology, incentives, advice, market development and political 'cover'.

With the maturing of state and private plantations, and the shift from national security to commercial, social and environmental objectives, the role of the state has had to be thoroughly reviewed. The changes have occurred in the FC's roles in plantation ownership, afforestation, promotion and regulation.

During recent years, economic policy has been the main driver to affect the FC's role in afforestation and plantation ownership. Governments now see the private sector as generally better placed to secure efficient management of commercial assets. The private sector usually has greater freedom in raising capital, greater flexibility in hiring and using staff, greater alertness to changes in markets, a more entrepreneurial approach to exploiting assets, and is less open to political interference in commercial decisions. Accordingly, emphasis has switched to encouraging planting by the private sector. The commission's holdings have been reduced to rationalize its structure in the interests of efficiency. The disinvestment has been modest – about 10 per cent of the area of state-owned plantations has been sold to the private sector – but more significant in the number of small plantations transferred. Experience was that a large state agency could not manage small plantations as cost-effectively as a private owner (often a neighbouring landowner was able to integrate the plantations bought from the FC with his existing estate).

However, in opposition to this pressure to transfer to the private sector has been the growing value of state plantations as recreational and environmental resources: the outcome of the measures the FC had taken to promote access, recreation and biodiversity. In the most recent review of the future of the Forestry Commission's plantations, public access for recreation was a major factor. Environmental NGOs and others also believed that the commission would be more effective in increasing the biodiversity of forests originally established as single-species plantations. In order to meet these concerns, the government introduced schemes to encourage private owners to provide public access and to increase the wildlife and recreational value of their plantations. These use a novel approach. Private landowners are invited to bid for cash in exchange for provision of public services, such as access, biodiversity or landscape enhancement. The bidder specifies the amount he wants. The FC judges which bids offer the best value for money. So far, these have had only modest effect. Private owners may prefer the freedom to manage and exploit their plantations to meet their own objectives, unimpeded by the wishes of the public and environmental pressure groups. But it is, of course, a long-term business.

The FC's roles as promoter and regulator have also evolved, partly in response to the changing demands on the plantations, and partly reflecting the new economic policy's focus on impartial regulation. The radical changes in the grant schemes for planting, managing and restoring woods illustrates how the FC has moved from being a promoter of commercial forestry to a role as an agency in encouraging the development of forestry to meet social and environmental objectives. Being at the centre of a wide-ranging con-sultation and approval process for felling and planting trees, it has become a regulator with the duty of balancing and reconciling conflicting interests and preventing environmental damage. Where it judges it necessary, it can require an environmental assessment. And it has the power, if not satisfied, to refuse approval to plant.

This change in the FC's role seems to have been welcomed. However, an intriguing recent development in the perception of the role of the state in forest regulation is the promotion of voluntary standards by the World Wide Fund for Nature (WWF) through the Forest Stewardship Council (FSC). The FSC aims to protect forests through an accreditation process that, after inspection, confers a badge of responsible forest management. This badge bestows an advantage in markets where environmentally concerned buyers wish to be assured that their purchases are not harming forests. Although the main targets are countries where national forests are suffering rapid rates of deforestation, the scheme is open to forest owners everywhere. In the UK, the FC, private owners and NGOs have worked with the FSC to produce the Woodland Assurance Standard, which is based on the FC's National Forestry Standard, extended to cover more aspects of forest management. The FC and some private owners have signed up.

Those in favour of this initiative see it as offering greater assurance of responsible forest management. In the view of the author, it is unnecessary: UK forests are not at risk, deforestation is not a problem, and the National Forestry Standard gives an assurance that forests are being well managed. However, while there may be differing views as to its future value, there is no doubt that it introduces more complexity and uncertainty within forest management. Capital markets naturally prefer simplicity and certainty; so its effect on plantation values is likely to be adverse. This is the point of relevance when judging the prospects for privatization.

The Woodland Assurance Standard may, in time, come to take over some of the FC's duties in forestry regulation. For the time being, the FC remains as the industry's regulator. Its role has been reflected through a change in its structure.

Balancing acts: Reconciling public-policy objectives and private-sector investment

The changes in the structure of the FC came as a result of the government's review in 1994 of the case for privatizing its plantations. This review illustrates how it sought to reconcile policy objectives for forestry with its economic policies favouring competition and the private sector. The review was not published; but the government consulted widely, and it is possible to identify the key considerations.

The principle options for the state-owned plantations were:

* sale, singly or as a whole;
* incorporation as one or more going concerns with a view to a flotation;
* creation of a state agency charged with management to meet commercial, social and environmental objectives.

Other options that were in the air at the time included sale and leaseback to the FC of plantations and selling long-term cutting rights with the land remaining in state ownership. Neither of these seems to offer any particular advantages over the main options to compensate for the additional complexities and uncertainties that they would introduce into forest management.

Sale of all the plantations to one or a few buyers would be problematic. Because of the sheer scale of the transaction, there would be few bidders. It would also worry the wood-processing industry: it preferred to deal with the supplier it knew rather than face one or a few new powerful players in the wood market. Some companies considered that they might have to bid themselves to avoid losing security of wood supply – for them, an unwelcome tying-up of capital they could use more profitably developing their businesses. Since they had a direct interest in a sustainable wood supply and, in some cases, experience of owning and managing forests in other countries, they might have become responsible forest owners in the UK. But they were not keen.

Sale of plantations singly or in small groups would be more feasible. The market was already developed and the consequences for wood supply were not threatening. However, it would obviously take a very long time to sell 1 million hectares.

Incorporation presented practical problems. As the FC's commercial operations were mixed up not only with unremunerative recreational and biodiversity projects, but also with its departmental, regulatory and promotional activities, it was almost impossible to assess the potential profitability of a stand-alone forestry corporation (or two or three of them). The financial information needed for a stock market valuation as a preliminary to a flotation – annual accounts in a commercial format for three years – was not available. As experience in Sweden, New Zealand and South Africa shows, it takes several years to extract a commercial forestry corporation out of a government department, with their different accounting systems, objectives, staff structures and corporate cultures. The new corporation must then operate successfully for several years to build up a track record. The low rate of return on commercial forestry suggested that this would be difficult. The corporation would depend upon subsidies being continued indefinitely.

The realistic timescale for major improvements in large organizations is long, not as long as the timescale for creating new forests, but far longer than the pressures of politics usually favour. In addition, as noted below, privatization was very controversial; it was doubtful if a project over, say, ten years could have been sustained.

The opposition to privatization came from forest users, the environmental NGOs, private owners, as well as the wood-based industry. There was much concern about the likelihood of loss of access, highly valued in densely populated areas. Furthermore, the FC had made much more progress in increasing the biodiversity of plantations. This was partly because they were older and, thus, easier to diversify at the stage of felling and regeneration through changing species, age structure and layout. But the Forestry Commission had also explicitly adopted wildlife and landscape enhancement as targets and was evidently taking them seriously, whereas the environmental movement feared that the profit motive would dominate private-sector management. It expected it to be easier to apply pressure on a state agency than on a large number of private owners. The wood processing industry preferred the devil it knew for the reasons set out above. There was therefore a large political opposition to privatization, set against the practical difficulties of a wholesale disposal of the state-owned plantations.

The outcome of the review and subsequent developments show the government achieving some of its broader economic objectives, while main-taining the provision of non-marketed benefits:

*   Regulation was separated from forest operations within the FC by setting up separate agencies – the Forestry Authority as regulator and the Forest Enterprise – with clearly distinguished roles and objectives.
*   Efficiency was pursued through the setting of financial and economic objectives for the Forest Enterprise, as the agency charged with managing the state forest estate.
*   The programme of rationalization through the sale of plantations that were costly to manage because of their size and/or location, and not highly valued for other benefits, such as recreation or biodiversity, was extended.
*   A new scheme to encourage private owners to provide access was intro-duced. This complemented the earlier changes in the grants for planting to promote broadleaves, native woodlands, diversification of species and greater attention to landscape impacts.
*   The reorganization of the Forestry Commission was followed by a review of its structure that led to a reduction in its staff numbers and reduced costs.

To date, this accommodation seems to have been accepted as a workable compromise between the drive for increased efficiency and the public interest in access, landscape and biodiversity. It is notable for what it achieved, what it might achieve and what it rejected.

Regulation is effectively separated from management. Privatization is proceeding in a fashion that does not threaten other benefits. Efficiency has been improved within the FC. These are valuable achievements.

The private sector has responded to the changed grant system by planting much more diverse woods. Public access has benefited from the access agreements and creation of community woodlands. The uniform plantations are being diversified as they are regenerated through the grant schemes. These are open-ended processes, already with some achievements to their credit and with the possibility of making substantial improvements if sustained over the long term.

But the bulk of state-owned plantations remain in public ownership. The sales programme is too limited to produce a major change. It could be expanded in the future. An obvious option is to transfer production forests to some of the large wood-processing companies. As noted above, they have experience of forest management and an interest in a secure wood supply. Given their access to finance, and their wish to maintain a reputation for environmental friendliness, it seems likely that they would be competent and responsible managers of forests.

Another possibility is to contract out management of plantations or groups of plantations. However, this achieves little as the assets remain in state ownership and the private-sector managers are not, in general, more competent than the Forest Enterprise. It suffers from a major disadvantage where the forest is to be managed for multiple benefits. The management contract then becomes complex in an attempt to specify in advance how, for example, production is to be traded against biodiversity over a long period. This is particularly so for the task of developing an even-aged plantation into a diverse forest, which requires a management plan executed over several decades. In such cases, short-term management contracts (e.g. five years) are not long enough to show results; and a very long contract defeats the object of using the private sector to increase competition. In the UK, this approach is seen as bringing more problems than it solves. The FC has, instead, adopted a culture in which its forest managers are encouraged to seek out cost-effective opportunities to diversify plantations. In this respect, it is still a pioneer.

Guidance

From the UK experience we can extract some useful pointers on:

- assessing the scale, the costs and the benefits of privatization;
- the merits of the options for wholesale privatization;
- ways of increasing private-sector participation short of total privatization.

Assessing the scale, costs and benefits

Two factors are of major importance in the assessment of potential benefits and costs:

1   *The intrinsic profitability of commercial forestry.* If the rate of return is high, privatization has a much better chance of success. If it is lower than the usual rate expected from long-term investments, the interest of private sector will be small and may be only from those wishing to asset-strip, with a quick exit. The government may have to promise to keep paying subsidies, promises that are usually discounted. The best approach will then be to seek out potential owners with a direct interest in wood production and other plantation products and services.
2   *The importance of the non-marketed services provided by the plantations.* The more that these are valued, and the more diverse and widespread they are, the greater the difficulty and cost of ensuring that they are maintained. At some level, difficult to specify precisely, but nonetheless clearly there, these costs will exceed the benefits from privatization.

Within these parameters, the government should be clear about what it is aiming to achieve. Benefits from privatization may include:

- a one-off capital sum;
- reduced annual budgets for the forestry agency;
- higher returns from wood production;
- lower costs of operation;
- new markets for wood;
- new enterprises exploiting the plantations.

The costs may include:

- reduced provision of free social and environmental services;
- increased costs of supervision by the state forestry authority;
- increased public expenditure on subsidies;
- disruption to wood supply and the wood-based industries;
- public disquiet.

A broad assessment of these will show whether privatization is worth pursuing.

Options for wholesale privatization

The two main options are incorporation and sale of the plantations as one or a few lots.

Incorporation would take many years to effect. If the rate of return on plantations is low, the privatized corporation would (and would be seen to) depend upon continuing subsidies whose future level cannot be guaranteed. It would be an uphill struggle. Its strategy would have to be to move quickly into more profitable downstream businesses. This could be seen as either disrupting existing wood processing businesses or injecting some valuable competition into the market for wood products. A probable outcome is that existing businesses would seek to get more control of wood supply.

If the rate of return is high, the prospects for the privatized corporation would be better. It would even probably seek to expand by developing or taking over wood-processing businesses. Potential monopoly power may then be a problem that has to be addressed.

Sale of a large share of a country's wood production potential to one or a few new owners threatens the wood supplies of the processors. As above, a low rate of return and dependence on future levels of subsidy makes these companies reluctant to commit capital to acquiring plantations. They could, perhaps, be manoeuvred into becoming owners if faced with the unpalatable prospect of control of their essential raw material passing to potential competitors. They would then be paying for the security of supply currently provided free by the state.

In an environment where the rate of return is good, of course, this should not be a problem. Wood processors may then be keen to acquire their wood supply. Either way, as owners, they bring the advantage of a continuing interest in sustaining the resource.

Given a transfer to the private sector, the UK experience is that a system of controls on felling, planting and replanting, with grants to compensate for low returns and/or negative impacts on profitability, can ensure that private forests are managed responsibly and without environmental damage. The systems must be competently and honestly administered, with consultation with other affected interests. The costs of this type of system are substantial. Other interests may, nonetheless, prefer that plantations remain with a public agency that is more open to political pressure and is less profit motivated.

Merely maintaining the level of non-market services may not be good enough. The aim may be, as in the UK, for increased biodiversity and social services. While a system of controls, grants and consultation can ensure responsible forest management, moving private owners to the next stage – an imaginative development of plantations to increase public benefits – is more problematic. Discretionary grant schemes may supply the answer, or voluntary certification; but it is too soon to judge the effectiveness of these approaches. However, it is clear that privatization does present the forestry authority with new or increased problems of regulation and promotion. In contrast, a public agency with an approved mandate is a direct means of achieving these goals.

Other ways of increasing private-sector participation

From the UK experience, there are several ways of increasing private-sector involvement that can work well and do not raise awkward problems:

- *Contract out felling (harvesting) and restocking (regeneration).* Provided that operations are supervised competently, there is no significant loss and there can be benefits from exploiting the private sector's drive to increase efficiency. The FC found that the staff managing the contracts needed training. It was also necessary to encourage workers to set up as contractors through contracts tailored to their capacity or financial circumstances.
- *Privatize some nurseries.* These small-scale operations are essentially similar to those already operating in horticulture; therefore, skills may already be available in the private sector. Note, however, that the success of plantation establishment and restocking depends crucially upon the quality of the young transplants supplied from the nursery. Contracts for supply must therefore specify both quality and handling of the seedlings sent to the forest, and deliveries to the site need careful inspection.
- *Sell plantations that are costly to manage and not of much value to other users.* The process is to identify those with high costs of management (for example, because they are small, distant from the forester's base or have poor access) and low actual or potential value for social or environmental services. It may be necessary to develop a market and set reserve prices to get a good return from the sale. There is scope for corruption; so methods of sale need to be transparent (e.g. public auction using independent land agents).
- *The restructuring of the FC to separate regulation from forest management* has also been successful and is no longer contentious. The process is to set up agencies for regulation and forest management within the Ministry of Forestry, and give them clear roles and objectives. Experience in the UK is that there can be substantial gains in efficiency and effectiveness from this step on its own. It allows for a narrower focus on work objectives, removes distractions, and facilitates the setting of targets and measurement of performance. As a result, it is worth doing on its own account, even if the state-owned plantations are not being privatized.

Conclusions and ways forward

Privatization of state-owned plantations is likely to be a publicly contentious policy. Numerous interests can be affected: local people using the plantation; environmental NGOs; the forest-based industry worried about future contracts for supply; and the state forestry service, which will be reluctant to accept a reduced or changed role. The government should be clear about what it seeks to achieve and what it needs to avoid. Consultation with those affected can help to reduce adverse impacts (and opposition).

The private sector will usually have greater flexibility, better access to finance, fewer restrictions, more drive and clearer objectives – although these may conflict with public interests, such as continued use by local people and protection of wildlife. There are ways of reducing adverse impacts – through selection of plantations, specific incentive schemes and the regulations – but at a cost.

The broad conclusion of the review and subsequent public discussion is that purely commercial plantations can be managed at least as well, and probably more efficiently, by the private sector, but that plantations with important public interest aspects are safer at present with the state forest service. Measures to ensure that these public interests would be safeguarded by the private sector are quite well established. However, getting the private sector to develop plantations into diverse forests that supply increased social and environmental services is much more difficult. Until this is shown to be feasible, it seems unlikely that wholesale privatization of all state-owned plantations will be an option in the UK. Nonetheless, there are opportunities, short of total privatization, to use the private sector, which can be beneficial.

Notes

Introduction

1 In this book the term 'natural forest' includes forests of natural origin, whatever degree of disturbance they may have been subjected to.

Chapter I

1 Net decline was 12.5 million hectares per year owing to 3.6 million hectares per year natural expansion.
2 No use was specified for the remaining 26 per cent of the forest plantation area.
3 However, it should be stressed that the scientific evidence for such claims is often lacking (see Landell-Mills and Porras, 2002).
4 Plantation management may be labour intensive; but low unit labour costs keep the gross cost per cubic metre low. Or it may be labour non-intensive so that even high wage rates do not result in comparatively high unit costs of production.
5 Thirty-seven per cent were classed as 'other' or ownership was not specified.
6 Twenty-two per cent were classed as 'other' or ownership was not specified.
7 For example, West Coast Environmental Law (1999) in Canada speculated that privatization, or increased corporate control over British Columbia Crown lands, would reduce protection of important environmental values. It observed that companies were pushing for privatization as a means of avoiding public oversight and returning to the days of unrestricted logging. Rather than protecting public lands, it claimed that privatization would lead to the replacement of mandatory requirements that allowed public access, buffer fish streams, protect against land slide, etc. by voluntary rules that provided much less protection.

Chapter 3

1 Experience since the South Africa case study was prepared serves to highlight the challenges of tackling labour relations, land claims and other aspects of transfers of state forest plantations. Former Department of Water Affairs and Forestry (DWAF) workers who were not supported by a poorly implemented social plan have burned down plantations. The shifting of workers from category A plantations to categories B and C led to staffing levels three times more than was needed and to workers fearful of losing their jobs burning down those plantations as well, substantially reducing the area and value of assets that can be sold. The issue of underlying land claims has not been adequately addressed in the lease agreement for category A plantations, with the result that the use rights of lessees are being challenged. The timetable for transferring all plantations to the private sector has proved to be unrealistic, but was, nevertheless, the basis for DWAF's medium-term financial plan. As a result, there has been a substantial shortfall in forestry's budget allocation. The procedure for transferring B and C plantations cannot now be used because the Department for Land Affairs is insisting on a more detailed community consultation process. As a result of these problems, the government may be forced to continue to manage (or to outsource the management of) the category B and C plantations for some years.

Chapter 5

1 The terms 'forest plantations' and 'plantation forestry' are used throughout this chapter to refer to stands of trees planted for commercial wood production. They are referred to in this way to distinguish them from other types of plantations.

2 Used here to also include woodlands, unless where otherwise stated. The definition of forest used by Australia's National Forest Inventory is 'an area ... dominated by trees ... [with a] mature stand height exceeding 2m and ... crown cover of overstorey strata about equal to or greater than 20 per cent' (NFI, 1998).

3 Between 20–50 per cent crown cover (NFI, 1998).

4 Using 1999 data: the fifth highest rate globally (CoA, 2002).

5 These tenures commonly preclude industrial wood production.

6 Mechanisms by which the government intervenes to change market prospects for particular activities to overcome inherent disadvantages faced by particular industries. This is the case in the plantation sector, where the taxation settings used aim to overcome the disincentive to investment caused by the length of time before return from investment in plantations.

7 Information for this section has been drawn from the annual reports of relevant agencies/companies and from personal communication with

representatives of Australian Capital Territory (ACT) Forests, Forest Products Commission, Forestry South Africa, Forestry Tasmania, State Forests New South Wales (SFNSW), and Hancock Victorian Plantations.

8   'The so-called "13-month" rule ... meant that the investment management company could raise funds from investors in one financial year, and then take 13 months to provide the services that had been paid for. For timber plantation projects, it was a useful provision because it enabled the investment manager to raise the funds and then apply them to getting the plantation "in the ground" in a timely way that could be matched to the seasonal requirements of plantation establishment' (Australian Forest Growers, 2000, p10).

9   'The government removed the 13-month rule in cases where the prepaid expenditure was greater than the investor's income from the same activity for that tax year. This policy change became law on 30 June 2000. Withdrawal of the rule meant that investor funds had to be spent on services provided in the same tax year as they were paid to the investment manager' (Australian Forest Growers, 2000, p10).

10   The definition of farm forestry varies – sometimes including or excluding joint venture and leasing schemes. In this section, joint ventures and lease schemes are referred to as they take place on land still owned and, in many cases, occupied by farmers; however, it should be recognized that there is generally little active involvement of the landholder compared to other forms of farm forestry.

11   A combination of physical (farm size), business (priorities, market uncertainties, relative investment returns, access to capital), and human (management expertise, labour availability) factors limited tree-growing.

Chapter 7

1   That is, the demand for paper tends to expand more rapidly than the growth in income.

2   Some discussions modify the regional organization of Table 7.2 to more closely reflect management responsibilities. In this case, the north-east and south-west are sometimes lumped (North-east South-west State Forest Region NSSFR) as the regions of predominantly state-owned forest and as China's major sources of industrial wood. The objective of these state forests has been to support employment in the local mills. Their standing inventory declined because the procurement prices paid by the mills were too low to support reforestation. The Southern Collective Forest Region (SCFR, including the south-east, south central, and some of the south-west from Table 7.2) is the second most productive region. This region also suffered from procurement prices that are still too low to encourage reforestation without external financial injections. The North/North-West Farm Forest Region (NNFFR) covers the remainder

of the country. The western part of the NNFFR is a dry area that is not conducive to forests. The eastern part is a major agricultural area. The only significant forest interventions in this region have been government-supported shelterbelts. Agricultural and other rural policy reforms have affected the SCFR and the NNFFR; but they may have had less impact elsewhere. They have led to vigorous agroforestry activities that have significantly altered the landscape in the eastern part of the NNFFR. More recent reforms of the SFA and the state forest bureaus, including a few large-scale government plantations, may have had greater impact on the NSSFR.

3   Because the collective forests support a disproportionate share of young stands.

4   This includes commercial tree crops such as orchards that contain less standing timber volume per unit area and are almost totally concentrated on the collective lands.

5   For a more extensive discussion, see Hyde et al (2003).

6   This became the third component of the 'three-fix' policy for forestry: stabilizing the rights and ownerships of forests and mountains; identifying boundaries of household plots; and establishing a forest production responsibility system.

7   Some plantations are composed of horticultural species or bamboo, which typically grow to smaller mature volumes than the volumes of either coniferous or deciduous forests.

8   Each household must allocate a family member to look after its many forest plots, while only one or two guards were necessary to protect an entire forest under collective management.

9   Bamboo has no harvest quotas or shipment restrictions, and has much lower schedules of taxes and fees than timber.

10 Flooding in the Yangtze and Songhua River basins in 1999, and dust storms in northern China in 2000–2001.

11 Including almost 500 mammalian species, 1189 species of birds, more than 320 reptiles and 210 amphibians.

12 These include breeding endangered animals for sale, the development of zoos, the establishment of amusement parks, sales of mounted specimens of rare species, illegal cutting and cultivation of timber, poaching of wildlife, and uncontrolled ecotourism.

Chapter 8

1   See www.undp.org.

2   In many parts of the country, however, the complete legal process for declaring an area as forest land has yet to be completed after the required settlement of the rights of their pre-existing occupants, if any.

3   This section draws on Saigal (1998).

4   US$1 = 48 rupees, approximately.

5   While some of the area has no tree cover naturally (e.g. snow-covered peaks, cold and hot deserts and wetlands), some areas have lost tree cover due to degradation.
6   There are, however, disputes over the state ownership in some areas and in many areas the complete legal process of declaring an area as forest land has not been completed.
7   Adapted from Saigal et al (2002).
8   It must, however, be clarified that data from a few states was not available and these percentages have been calculated based on a total area of 69.64 million hectares.
9   Between 1981–1982 and 1985–1986, projects totalling 9.9 billion rupees were initiated in 14 states (MoEF, 1989, in Vira, 1995).
10 This section draws on Saigal (1998).
11 According to estimates, over 458 million metric tonnes of wet dung were being used annually as fuel. If this was used in agriculture fields, it could potentially fertilize 91 million hectares and increase food output by 45 million metric tonnes (Srivastava and Pant, 1979, in Pant, 1979).
12 These are known by different names in different states (e.g. Vana Samaraksha Samitis in Andhra Pradesh and Hill Resource Management Societies in Haryana) but are most commonly referred to as forest protection committees or FPCs.
13 Many tribal-dominated areas, excluding the north-eastern states, are included in Schedule V of the Constitution of India. These areas have special provisions for their administration.
14 The forest development agencies, however, need substantial strengthening in order for them to become truly representative of community interests. At present, these are dominated by the Forest Department. Another criticism of the National Afforestation Programme is that it imposes a uniform JFM structure on all the states and does not take into account progressive measures regarding FPC structure and functioning introduced in many states.

Chapter 9

1   Forests on land reserved under the South Island Landless Natives Act 1906 (SILNA), timber from land administered by the Department of Conservation and planted indigenous forest are exempt.
2   The make-up of the processing sector changes rapidly in New Zealand and so these figures are likely to date rapidly.
3   Throughout this chapter, the name New Zealand Forest Service (NZFS) is used to avoid confusion.
4   Equivalent to 5200ha.
5   Equivalent to 120,000ha.
6   Particularly tree-planting by small land owners, usually referred to as farm forestry.

7 Formed from Perpetual Forests Ltd – formerly the largest bond company in New Zealand.

8 In the NZFC, 7070 employees engaged in commercial forestry (both salary and wage workers) were reduced to 2770 (Birchfield and Grant, 1993, p77).

9 Under the decision, 'All wage workers would be offered a choice: one year's guaranteed work from 1 April [1987] with no promises and no redundancy thereafter or redundancy, with compensation based on the standard state sector formula, on 31 March' (Birchfield and Grant, 1993, p70, emphasis in original).

10 A valuation would effectively be obtained through the sale of the assets (Birchfield and Grant, 1993).

11 By now renamed the Forestry Corporation of New Zealand (FCNZ).

12 Under the Crown Forestry Rental Trust.

13 'the draft agreement provided that, if the Waitangi Tribunal recommended land be returned to Maori ownership, the Crown would transfer its ownership to the successful claimant – including the right to the rental from that time and then progressively resume control of the land as trees were clear-felled; compensate the successful claimants for the fact that the then existing crop was retained by someone else by paying 5 per cent of the value of the trees; and further compensate the successful claimant by paying such further proportion of the value of the tree crop as the Waitangi Tribunal recommended' (Birchfield and Grant, 1993, p180).

14 For example, if a CFL had a covenant requiring the holder to maintain an area for conservation purposes, the level of rent might be reduced accordingly.

15 In particular, planning for transport and infrastructure development and allocating funding for improving and upgrading roads required for harvesting plantations, especially the increased area of plantations established in a wide range of rural regions from the 1960s onwards (Inwood News Services, 2002d).

16 The East Coast Forestry Project (ECFP) established in 1992 is the only area in which the Crown is still involved in actively expanding the plantation estate in New Zealand through the allocation of grants. Initially, the ECFP aimed to plant 200,000ha of plantations over 28 years on eroding and erosion-prone land on the east coast of North Island. The project originally aimed both to reduce erosion and to provide social benefits in the form of employment and regional development opportunities, particularly for the Maori, in a region that has had one of the highest unemployment rates in New Zealand (Office of the Parliamentary Commissioner for the Environment, 1994; Cocklin and Wall, 1997). In 1999, the primary objective changed to achieving sustainable land management (MAF, 2002b).

References

ABARE (Australian Bureau of Agricultural and Resource Economics) (2001) *Australian Commodity Statistics 2001*, ABARE, Canberra

ACF (Australian Conservation Foundation) (1988) *The Wood and the Trees Revisited: ACF Responds to the Forest Industries*, ACF, Hawthorn, Victoria

Acuiti Legal (2002) *Tax Concessions Reintroduced for Plantation Forestry Industry*, www. ptaa.com.au/acuiti.html, accessed 30 April 2002

AFFA (Agriculture Fisheries and Forestry – Australia) (2002) Industry Development and Adjustment. Australia's Unique Forests, www.affa.gov.au/, accessed 31 July 2002

Afforestation Proprietary Limited (1926) *Timber Wealth, Something For All: A Large Return for a Small Investment*, Afforestation Proprietary Ltd, Melbourne

AFS Steering Committee (Australian Forestry Standard Steering Committee) (2002) 'An Australian forestry standard', www.forestrystandard.org.au/index.html, accessed 7 August 2002

Aldwell, P. H. B. (1984) *Some Social and Economic Implications of Large-scale Forestry in Waiapu County*, Forest Research Institute, New Zealand Forest Service, Rotorua

Alexander, F., Brittle, S., Ha, A. G., Leeson, T. and Riley, C. (2000) *Landcare and Farm Forestry: Providing a Basis for Better Resource Management on Australian farms*, ABARE report to the Natural Heritage Trust, Canberra

Anderson, J. (1997) *Minister Removes Export Controls on Wood Sourced from Plantations in Western Australia*, Media Release Department for Primary Industries and Energy, 22 September 1997

Anderson, R. S. and Huber, W. (1988) *The Hour of the Fox: Tropical Forests, The World Bank and Indigenous People in Central India*, Vistaar/Sage, New Delhi

Anon (1996) 'NZ government rushes to sell forests', *Green Left Weekly*, www.greenleft. org.au/back/1996/245/245p18.htm, accessed 28 November 2002

Anon (1998) 'Job cuts decimate New Zealand forestry industry', *World Socialist*, www. wsws.org/news/1998/nz-d23.shtml

Anon (2001) 'Gunns adds North Forest to its trees', *The Australian*, 20 March 2001

APEC (Albany Plantation Export Company) (2002) 'Plantation to port: The dawning of a new era', Advertising feature, *Albany Advertiser*, 7 March 2002

Australian Forest Growers (2000) 'Submission to the Senate Economics References Committee', Inquiry into Mass Marketed Tax Effective Schemes and Investor Protection, AFG, ACT, www.afg.asn.au, accessed 20 March 2002

Banerjee, A. K. (2000) 'Devolving forest management in Asia-Pacific countries', in Enters, T., Durst, P. B. and Victor, M. (eds) *Decentralisation and Devolution of Forest management in Asia and the Pacific*, RECOFTC Report No 18 and RAP Publication 2000/2001, Bangkok, Thailand

Bass, S., Dubois, P., Moura-Costa, P., Pinard, M., Tipper, R. and Wilson, C. (2000) 'Rural livelihoods and carbon management: An issues paper', UK DFID Forestry Research Programme Project R7374

Bass, S. and Hearne, R. (1997) *Private Sector Forestry: A Review of Instruments for Ensuring Sustainability*, Forestry and Land Use Series No 10, International Institute for Environment and Development (IIED), London

Bass, S., Thornber, K., Markopoulos, M., Roberts, S. and Grieg-Gran, M. (2001) Certification's Impacts on Forests, Stakeholders and Supply Chains, International Institute for Environment and Development (IIED), London

Beyer, M. (2002) 'Barriers cleared', *Western Australian Business News*, 11 April 2002, www.businessnews.com.au/news.cfm?function=show&NewsID=7884, accessed 30 April 2002

Bhati, U. N. (2002) ANU Forestry Log Market Report, www.sres.anu.edu.au/associated, accessed 3 August 2002

Binkley, C. (2002) Personal communication, Hancock Natural Resources Group, Boston

Birchfield, R. J. and Grant, I. F. (1993) *Out of the Woods: The Restructuring and Sale of New Zealand's State Forests*, GP Publications, Wellington

Buffier, B. (2002) *Environmental and Commercial Outcomes through Agroforestry: Policy and Investment Options*, RIRDC Publication 02/057, Rural Industries Research and Development Corporation, Canberra

Burns, K., Walker, D. and Hansard, A. (1999) *Forest Plantations on Cleared Agricultural Land in Australia: A Regional Economic Analysis,* ABARE Research Report 99.11, Canberra

BWG (Biodiversity Working Group) (1997) *Conserving China's Biodiversity: Report of BWG to the China Council for International Cooperation on Environment and Development*, CCICED, Beijing

Byron, N. and Boutland, A. (1987) 'Rethinking private forestry in Australia: Strategies to promote private timber production'. *Australian Forestry*, vol 50, no 4, 236–244

Byron, R. N. and Douglas, J. J. (1981) *Log Pricing in Australia*, BFE Press, Canberra

Cadman, T. (2002) *Australian FSC Stakeholders' Conference,* www.certifiedforests.org.au/, group email 27 March 2002, site accessed 6 August 2002

CAFLU (Chinese Agriculture and Forestry Labor Union) (1997) 'Developing agricultural economy to absorb surplus labor in forestry areas', *Forestry Economics*, no 1

Cai, H. H. (1999) 'Why social forestry in Vietnam? A trainer's perspective', *Social Forestry Training Network in Vietnam*, no 1, October 1999 Social Forestry Support Programme Helvetas, Hanoi. In Enters, T., Durst, P. B. and Victor, M. (eds) *Decentralisation and Devolution of Forest Management in Asia and the Pacific,* RECOFTC Report No 18 and RAP Publication 2000/2001, Bangkok, Thailand

Campbell N., Storey, T. and Taylor, C. (2001) 'The forestry taxation regime', *New Zealand Journal of Forestry,* vol 46(3), pp34–35

Carron, L. T. (1985) *A History of Forestry in Australia*, The Australian National University Press, Canberra

Carron, L. T. (1990) 'A history of plantation policy in Australia', in Dargavel, J. and Semple, N. (eds) *Prospects for Australian Forest Plantations*, Centre for Resource and Environmental Studies, Canberra, pp11–23

Catterall, C. P. (2000) 'Wildlife biodiversity challenges for tropical rainforest plantations', in Snell, A. and Vise, S. (eds) *Opportunities for the New Millennium*, Proceedings of the Australian Forest Growers Biennial Conference, Cairns, pp191–195

CCTV (China Central Television) (2000) 'Growing forests on report cards', 30 May

CFIC (China Forestry Information Center) (1998) *Development of China's Wood Industry*, Chinese Academy of Forestry, Beijing

Chalmers, R (2001) 'Innovative restructuring of state's Eastern Cape forestry assets will help poor – Seeing the wood and the trees', *Business Day*, August 2001

China Customs Office (1999–2001) *China Customs Office Statistical Yearbook*, China Customs Office Press, Beijing

China National Bureau of Statistics (2002) *Yearbook of Statistics 2002*, CNBS, Beijing

Clark, J. (2002) *Australian Production of Wood and Wood Products in 1999/2000 and 1989/1990 Disaggregated by Wood Source*, August 2002, Unpublished manuscript for the Centre for Resource and Environmental Studies, Australian National University, Australia

Clarke, M. (1999) 'Devolving forest ownership through privatization in New Zealand', *Unasylva*, no 199, vol 50(4), pp35–44

Clarke, M. (2000) 'Devolving forest ownership in New Zealand: Processes, issues and outcomes', in Enters, T., Durst, P. B. and Victor, M. (eds) *Decentralisation and Devolution of Forest Management in Asia and the Pacific*, RECOFTC Report No 18 and RAP Publication 2000/2001, Bangkok, Thailand

CMIE (Centre for Monitoring Indian Economy) (2003) *National Income Statistics, January 2003*, Centre for Monitoring Indian Economy Pvt. Ltd, New Delhi

Cochrane, P. and Gerritsen, R. (1990) *The Public Policy Implications of Eucalypt Plantation Establishment in Australia: An Introductory Survey*, Discussion Paper No 20, Public Policy Program, Australian National University, Canberra

Cocklin, C. and Wall, M. (1997) 'Contested rural futures: New Zealand's East Coast Forestry Project', *Journal of Rural Studies*, vol 13(2), pp149–162

CoA (Commonwealth of Australia) (1992) *National Forest Policy Statement*, AGPS, Canberra, www.rfa.gov.au, accessed 4 August 2002

CoA (1996) *State of the Environment Australia 1996*, CSIRO Publishing, Melbourne

CoA (1999) *Regional Australia Summit*, www.dotrs.gov.au/regional/summit/, accessed 4 August 2002

CoA (2000) *Regional Forest Agreements*, www.rfa.gov.au, accessed 31 July 2002

CoA (2001) *The Farm Forestry Program (FFP)*, www.affa.gov.au/docs/1_nrm/nht_landcare/nht/ffp-summary.html, accessed 30 July 2002

CoA (2002) *State of the Environment Australia 2001*, CSIRO Publishing, Melbourne

CONAF (Corporación Nacional Forestal) (2000) *Gerencia de Desarrollo y Fomento Forestal: Gestión Período 1997–1997*, CONAF. Santiago

Cox, O., Horgan, G. and Maplesden, F. (1993) *The New Zealand Forestry Sector in 1993*, FRI Bulletin No 185, New Zealand Forest Research Institute, New Zealand

CRC–SPF (Cooperative Research Centre for Sustainable Production Forestry) (2002) www.forestry.crc.org.au, accessed 29 July 2002

CSE (Centre for Science and Environment) (1982) *The State of India's Environment, 1982: A Citizen's Report*, CSE, New Delhi

Danodoran, A. (n.d.) Tree Plantations and Biodiversity in the Indian Context. Indian Institute of Plantation Management, Bangalore (Mimeo)

Dargavel, J. (1995) *Fashioning Australia's Forests*, Oxford University Press, Melbourne

Dargavel, J., Guijt, I., Kanowski, P., Race, D. and Proctor, W. (1998) *Australia: Settlement, Conflicts and Agreements*, Discussion paper for IIED, London

Department of Natural Resources and Environment – Victoria (2002) *Forestry Rights and Carbon Sequestration Rights*, www.nre.vic.gov.au, accessed 27 July 2002

DETR (UK Department of the Environment, Transport and the Regions) (2000) *Climate Change: The UK Programme*, HMSO, Cm4913, London

DFID (UK Department for International Development) (2002) *Southern Africa: Social Appraisal for Project Document 'Making Forest Markets Work for the Poor'*, Internal Report

Dlomo, M. and Pitcher, M. (2002) 'Changing Ownership and Management of State Forest Plantations: South Africa's experience'. Draft study submitted to the International Conference, Changing Ownership and Management of State Forest Plantations: Issues, Approaches, Implications. Cape Town, South Africa, 6–3 November 2002

DLWC (Department of Land and Water Conservation) (2001) *Caring for our Natural Resources: Plantations*, www.dlwc.nsw.gov.au/care/veg/plantation/, accessed 29 July 2002

Dogra, B. (1985) 'The World Bank vs the people of Bastar', *The Ecologist*, vol 15(1/2), pp44–49

Dutta, M. and Adhikari, M. (1991) *Sal Leaf Plate Making in West Bengal: A Case Study of Cottage Industry in Sabalmara, West Midnapore District*, Indian Institute of Bio-social Research and Development, Calcutta

DWAF (Department of Water Affairs and Forestry) (1996) 'White Paper on Sustainable Forestry Development in South Africa', Government Gazette, South Africa

DWAF (1998) *National Forests Act, No 84*, Government Gazette, South Africa

DWAF (2003) *State of the Forests Report 2003*, DWAF, Pretoria

Dwivedi, B. N. (1993) 'Wasteland development in India', in Rawat, A. S. (ed) *Indian Forestry: A Perspective*, Indus Publishing Company, New Delhi.

Dwyer Leslie Pty Ltd and Powell, R. A. (1995) Oberon: Rural Community Development Study Final Report. December 1995. State Forests of New South Wales, Sydney, 125pp

Emborg, J. and Larsen, J. B. (1999) 'How to develop plantations into forests – in order to achieve stability and functional flexibility? A north European perspective', in Barros, S. and Campdonici, M. I. (eds) *Proceedings of the International Experts Meeting on the Role of Planted Forests in Sustainable Forest Management*, Corporacion National Forestal, Santiago de Chile

Enters, T., Durst, P. B. and Victor, M. (eds) (2000) *Decentralisation and Devolution of Forest Management in Asia and the Pacific*, RECOFTC Report No 18 and RAP Publication 2000/2001, Bangkok, Thailand

European Economic and Social Committee (2002) 'Section for Agriculture, Rural Development and Environment', Opinion on the eastward enlargement of the European Union and the forestry sector, *Official Journal of the European Communities*, C 149/51, 21 June 2002

FAFPIC (Forestry and Forest Products Industry Council) (1987) *Forest Industries Growth Plan: A Submission to the Australian Government*, FAFPIC, Melbourne

FAO (United Nations Food and Agriculture Organization) (2001a) *Global Forest Resources Assessment 2000*, FAO Forestry Paper 140, FAO, Rome

FAO (2001b) *Biological Sustainability of Productivity in Successive Rotations*, Forest Plantation Thematic Paper Series, Working Paper FP/2 by J. Evans, March 2001

FAO (2001c) *Plantations and Wood Energy*, Forest Plantation Thematic Paper Series, Working Paper FP/5 by D. J. Mead, March 2001

FAO (2001d) *Financial and Other Incentives for Plantation Establishment*, Forest Plantation Thematic Paper Series, Working Paper FP/8, J. Williams, March 2001

FAO (2001e) *The Impact of Forest Policies and Legislation on Forest Plantations*, Forest Plantation Thematic Paper Series, Working Paper FP/9 by C. J. K. Perley, March 2001

FAO (2001f) *Forestry Out-grower Schemes: A Global View*, Forest Plantation Thematic Paper Series, Working Paper FP/11 by D. Race and H. Desmond, March 2001

FAO (2001g) *Plantations and Greenhouse Gas Mitigation: A Short Review*, Forest Plantation Thematic Paper Series, Working Paper FP/12, by P. Moura-Costa and L. Auckland, March 2001

FAO (2001h) *Future Production from Forest Plantations*, Forest Plantation Thematic Paper Series, Working Paper FP/13, by C. Brown, March 2001

Fisher, R. J. (2000) 'Decentralisation and devolution in forest management: A conceptual overview', in Enters, T., Durst, P. B. and Victor, M. (eds) *Decentralisation and Devolution of Forest Management in Asia and the Pacific*, RECOFTC Report No 18 and RAP Publication 2000/2001, Bangkok, Thailand

Forest Stewardship Council (2000) FSC-STD-01-001. 'FCC Principles and Criteria for Forest Stewardship', FSC, Bonn

Forestry Commission (1998) *England Forestry Strategy*, Forestry Commission, National Office for England, Cambridge, UK

Forestry Commission (2000a) *Scotland's Forests*, Forestry Commission, National Office for Scotland, Edinburgh, UK

Forestry Commission (2000b) *Wales Woodland Strategy*, Forestry Commission, National Office for Wales, Aberystwyth, UK

Forestry Commission (2001) *Forestry Statistics, 2001*, Forestry Commission, Edinburgh, UK, November 2001

Forestry Commission (2002) *Indicators of Sustainable Forestry*, Forestry Commission, Edinburgh, UK

Forestry Joint Venture Working Group (1991) *Report of the Forestry Joint Venture Working Group July 1991*, Ministry of Forestry, New Zealand

FPB (Forest Practices Board) (2000) *Forest Practices Code 2000*, www.fpb.tas.gov.au/fpb/docs/code_contents.htm, accessed 28 July 2002

FPCom (Forest Products Commission) (2001) *Forest Products Commission 2000–2001 Annual Report*, FPC, Western Australia

FSC (Forest Stewardship Council) (2000) *FSC Principles and Criteria*, (revised February 2000), Forest Stewardship Council, Oaxaca

FSI (Forest Survey of India) (1999) *The State of Forest Report*, FSI, Ministry of Environment and Forests, GoI, Dehradun

FWPRDC (Forest and Wood Products Research and Development Corporation) (2002) 'About FWPRDC', www.fwprdc.org.au, accessed 29 July 2002

Garforth, M and Dudley, N. (2003) *Forest Renaissance: The Role of State Forestry in Britain, 1919–2050: A Discussion Paper*, Forestry Commission, Edinburgh, UK

Garforth, M. and Thornber, K. (2003) *Impacts of Certification on UK Forests*, Unpublished report for the Forestry Commission

Ghose, A. (2001) Status of JFM in Rajasthan. Draft paper prepared for the book on Joint Forest Management being prepared by the Commonwealth Forestry Association in collaboration with Winrock International India and Ford Foundation, New Delhi

Gluckman, R. (2000) 'The desert storm', *Asiaweek*, 13 October, pp36–40

GoI (Government of India) (1894) 'Circular No 22-F', 19 October, Department of Revenue and Agriculture, GoI, Calcutta

GoI (1952) *National Forest Policy Resolution*, 12 May, Ministry of Food and Agriculture, GoI, New Delhi

GoI (1976) *Report of the National Commission on Agriculture: Forestry, vol IX*, Ministry of Agriculture and Irrigation, GoI, New Delhi

GoI (1984) *Report of the Committee for Review of Rights and Concessions in the Forest Areas of India,* Ministry of Agriculture, GoI, New Delhi

GoI (1988) *National Forest Policy Resolution,* Ministry of Environment and Forests, GoI, New Delhi

GoI (1990) *Report of the Team of Experts to Review the Functioning of State Forest Development Corporations under the State Governments/Union Territories,* Ministry of Environment and Forests, GoI, New Delhi

GoI (1999) *National Forestry Action Programme – India,* Ministry of Environment and Forests, GoI, New Delhi

GoI (2002a) *Economic Survey 2001–2002,* Ministry of Finance, GoI, New Delhi

GoI (2002b) *Joint Forest Management: A Decade of Partnership,* Ministry of Environment and Forests, GoI, New Delhi

Grants, T. C. (1989) *History of Forestry in New South Wales, 1788 to 1988,* D. Grant, Sydney

Gray, I. and Lawrence, G. (2001) 'The global misfortune of regional Australia', in Gray, I. and Lawrence, G. (eds) *A Future for Regional Australia: Escaping Global Misfortune,* Cambridge University Press, Cambridge, pp 1–15

Grebner, D. L. and Amacher, G. S. (2000) 'The impacts of deregulation and privatization on cost efficiency in New Zealand's forest industry', *Forest Science,* vol 46(1), pp40–51

Grundy, D. (2002) 'Changing Ownership and Management of State Forest Plantations: The United Kingdom's Experience'. Draft Study Submitted to the International Conference, Changing Ownership and Management of State Forest Plantations: Issues, Approaches, Implications. Cape Town, South Africa, 6–8 November 2002

Gunningham, N. and Sinclair, D. (2001) 'Voluntary approaches to environmental protection: Lessons from the mining and forestry sectors?', Paper commissioned by the OECD Environment Directorate

Hall, O. F. (1997) 'New Zealand's privatization of forest lands: Policy Lessons for the United States and elsewhere?', *Forest Science,* vol 43(2), pp181–193

Harkness, J. (1998) 'Recent trends in forestry and conservation of biodiversity in China', *The China Quarterly,* pp911–934

Harwood, C. and Bush, D. (2002) *Breeding Trees for the Low Rainfall Regions of Southern Australia,* RIRDC, Canberra, www.rirdc.gov.au/

Hawes, P. and Memon, P.A. (1998) 'Prospects for sustainable management of indigenous forests on private land in New Zealand', *Journal of Environmental Management,* vol 52, pp113–130

Healy, B. (1982) *A Hundred Million Trees: The story of NZ Forest Products Ltd,* Hodder and Stoughton, Auckland

Hobley, M. (1996) *Participatory Forestry: The Process of Change in India and Nepal,* Rural Development Forestry Study Guide No 3, Overseas Development Institute, London

Hobley, M. (2002) *Responding to the Challenge: Forest Enterprise and Community Involvement in Scotland,* Report for the Forestry Commission's Forests and People Advisory Panel, Forestry Commission, Edinburgh

Horton, M. (1995) *Clearcut: Forestry in New Zealand,* Campaign against Foreign Control of Aotearoa, Christchurch, Unpublished manuscript

Hurley, P. J. (1986) *Economic Incentives for Private Softwood Plantations in Victoria,* MSc thesis in Forest Science, Faculty of Agriculture and Forestry, University of Melbourne, Parkville, Victoria

Hyde, W., Belcher, B., Xu, J. (eds) (2002) *China's Reforms and Its Forestry Sector: Lessons for Global Forest Policy*, Unpublished manuscript available from Centre for International Forestry Research, Bogor, Indonesia

Hyde, W., Xu, J. and Belcher, B. (2002) 'Introduction', in Hyde, W., Belcher, B. Xu, J. (eds) *China's Reforms and Its Forestry Sector: Lessons for Global Forest Policy*, Unpublished manuscript available from Centre for International Forestry Research, Bogor, Indonesia

IASSI (Indian Association of Social Science Institutions) (1991) *Land Reform and Rural Change*, IASSI, New Delhi

ICFRE (Indian Council of Forest Research Education and Training) (2000) *Forestry Statistics India – 2000*, ICFRE, Dehradun

IIED (International Institute for Environment and Development) (1996) *Towards a Sustainable Paper Cycle*, World Business Council for Sustainable Development and IIED, London

Indufor and Eco (2001) *Implications of Land Restitution for Achieving World Bank/WWF Alliance Targets in Eastern Europe and the Central Asian Region: Main Report*, Report for the World Bank/WWF Alliance, WWF International Danube–Carpathian Programme Office, Vienna

INFOR (Instituto Forestal) (1997) 'National de Estudisticas, vicenso National Agropecuario 1997', Santiage de Chile

INFOR (2001) 'Estadisticas Forestales 2000', Santiago de Chile

Inglis, D. J., Higgs, M. L. and Stringer, D. R. (1985) 'Tasmania: A small competitor in a sea of softwood', in Meads, D. J. and Ellis, R. C. (eds) *Forestry – Satisfying National and Regional Needs,* Papers presented to the Second Joint Conference of the Institute of Foresters of Australia and New Zealand Institute of Foresters Inc, 20–24 May 1985, University of Canterbury, New Zealand, pp21–24

Inwood News Services (2002a) 'New Zealand wants FSC', News report 22 October 2002, New Zealand Forestry Industry Information Centre, www.nzforestry.co.nz, accessed 16 January 2003

Inwood News Services (2002b) 'CHH boss joins call to action', News report 15 October 2002, New Zealand Forestry Industry Information Centre, www.nzforestry.co.nz, accessed 16 January 2003

Inwood News Services (2002c) 'McFadgen Calls for urgent changes', News report 14 October 2002, New Zealand Forestry Industry Information Centre, www.nzforestry.co.nz, accessed 16 January 2003

Inwood News Services (2002d) 'Roading rescue gets closer', News report 11 October 2002, New Zealand Forestry Industry Information Centre, www.nzforestry.co.nz, accessed 16 January 2003

Inwood News Services (2003a) 'Liddell warns government on policies', News report 15 January 2003, New Zealand Forestry Industry Information Centre, www.nzforestry.co.nz, accessed 16 January 2003

Inwood News Services (2003b) 'UBS buys 8% of FCF's forest', News report 16 January 2003, New Zealand Forestry Industry Information Centre, www.nzforestry.co.nz, accessed 16 January 2003

IPCC (Intergovernmental Panel on Climate Change) (2000) *Land Use, Land-Use Change and Forestry*, Special Report of the Intergovernmental Panel on Climate Change, Cambridge University Press, UK

Jeanrenaud, S. (2001) *Communities and Forest Management in Western Europe*, IUCN, Gland, Switzerland

Kadekodi, G. K. (2001) *Notes on Macro Economics of Biodiversity in India*, Paper prepared for the National Biodiversity Strategy and Action Plan

Kanowski, P. J. (1997) 'Plantation forestry for the 21st century', *Proceedings, XI World Forestry Congress*, Antalya, Turkey, October, vol 3, pp23–34

Keenan, R. and Grant, A. (2000) *Implications of the Kyoto Protocol for Plantation Development*, Proceedings of the International Conference on Plantation Development, 7–9 November, Department of Environment and Natural Resources, Republic of Philippines, International Tropical Timber Organization and Food and Agriculture Organization, Manila, the Philippines

Keey, R. B. (2001) *Submission of Professor Emeritus Roger Brian Keey FRSNZ in Support of the Petition of Ross Carter and 7447 Others 1999/0170 to the House of Representatives*, Presented to the Local Government and Environment Select Committee of the House, September 2001, New Zealand

Kelly, G. and Lymon, K. (2000) *To Trees, or Not to Trees? An Assessment of the Social Impact of the Plantation Industry on the Shire of Plantagenet*, Curtin University of Technology, WA

Kelsey, J. (2002) 'New Zealand in the Asia Pacific: An economic and social overview', *2001 ICEM-A/P Regional Conference 'Solidarity with Pacific Region'*, Country Report, www.icemap.org/rc01-key.htm, accessed 28 November 2002

Khare, A., Sarin, M., Saxena, M. C., Palit, S., Bathla, S., Vania, F. and Satyanarayana, M. (2000) *Joint Forest Management: Policy, Practice and Prospects*, Policy that Works for Forest and People Series No 3, World Wide Fund for Nature – India, New Delhi, and International Institute for Environment and Development, London

Kirkland, A. and Berg, P. (1997) *A Century of State-honed Enterprise: 100 Years of State Plantation Forestry in New Zealand*, Profile Books, Auckland

Knock, P. (1993) *The 1987 Restructuring of Government Departments and Subsequent Incorporation and Privatisation of Forestry in New Zealand and the Effects on Access and Recreational Opportunities: The Organisation and Policies of Agencies Dealing with Forest Recreation in the USA*, Nuffield Farming Scholarship Trust, Uckfield, East Sussex, UK

Kun, Z. (2000) 'Issues relating to the reform of forest management in China', in Enters, T., Durst, P .B. and Victor, M. (eds) *Decentralisation and Devolution of Forest Management in Asia and the Pacific*, RECOFTC Report No 18 and RAP Publication 2000/2001, Bangkok, Thailand

Kunge, J. (2000) 'Rising giant enters the world', *Financial Times' Survey*, 13 November

Lai, C. K., Catacutan, D. and Mercado, A. R. (2000) 'Decentralising natural resources management: Emerging lessons from ICRAF collaboration in southeast Asia', in Enters, T., Durst, P. B. and Victor, M. (eds) (2000) *Decentralisation and Devolution of Forest Management in Asia and the Pacific*, RECOFTC Report No 18 and RAP Publication 2000/2001, Bangkok, Thailand

Lal, J. B. (1992) *India's Forests: Myth and Reality*, Natraj Publishers, Dehradun

Landell-Mills, N. and Porras, T. I. (2002) *Silver Bullet or Fools' Gold? A Global Review of Markets for Forest Environmental Services and Their Impact on the Poor*, Instruments for Sustainable Private-sector Forestry Series, IIED, London

Landell-Mills, N. and Ford, J. (1999) *Privatising Sustainable Forestry: A Global Review of Trends and Challenges*, IIED, London

Le Heron, R. B. and Roche, M. M. (1985) 'Expanding exotic forestry and the extension of a competing land use for rural land in New Zealand', *Journal of Rural Studies*, vol 1(3), pp211–229

LHA Management Consultants (2001) *Dynamics of the SA Hardwood Pole Market*, January 2001

Li J., Fanwen, K., Naihui, H. and Ross, L. (1988) 'Price and policy: The keys to revamping China's forestry resources', in Repetto, R. and Gillis, M. (eds) *Public Policies and the Misuse of Forest Resources*, Cambridge University Press, Cambridge

Lindenmayer, D. (2000) *Islands of Bush in a Sea of Pines: Summary of Studies from the Tumut Fragmentation Experiment*, Land and Water Australia Report PR000342, www.lwa.gov.au/products.asp, accessed 4 August 2002

Lindenmayer, D. B., Hobbs, R .J. and Salt, D. (2002) 'Plantation forests and biodiversity conservation', *Proceedings Australian Forest Plantations 2002*, BRS/ANU, www. affa.gov.au>Scientific Advice>Forest and Vegetation Sciences>Australian Forest Plantations Conference August 2002, accessed 11 September 2002

Lindsay,J.M. (2000) 'Creating legal space for community-based management:Principles and dilemmas', in Enters, T., Durst, P. B. and Victor, M. (eds) *Decentralisation and Devolution of Forest Management in Asia and the Pacific*, RECOFTC Report No 18 and RAP Publication 2000/2001, Bangkok, Thailand

Liu, D. and Edmunds, D. (2002) 'Devolution as a means of expanding local forest management in South China: Lessons from the last 20 years', in Hyde, W., Belcher, B. and Xu, J. (eds) *China's Reforms and Its Forestry Sector: Lessons for Global Forest Policy*, Centre for International Forestry Research, Bogor, Indonesia, Unpublished manuscript

Liu, J., Xiao, W. and Landell-Mills, N. (2002) 'Taxes and fees in the South China Collective Forest Region', in Hyde, W., Belcher, B. and Xu, J. (eds) *China's Reforms and Its Forestry Sector: Lessons for Global Forest Policy*, Centre for International Forestry Research, Bogor, Indonesia, Unpublished manuscript

Lu Wenming, N. Landell-Mills, Liu Jinlong, Xu Jintao and Liu Can (2002) *Getting the Private Sector to Work for the Public Good: Instruments for Private Sector Sustainable Forestry in China*, Instruments for Sustainable Private Sector Forestry Series, International Institute for Environment and Development, London

MacKinnon, J. R., Cheng, M. and Melville, C. (1995) A Biodiversity Review of China, WWF-Hong Kong, Hong Kong

MAF (Ministry of Agriculture and Forestry) (2001) *A National Exotic Forest Description as at 1 April 2000*, Edition 17, May 2001, www.maf.govt.nz/statistics/ primaryindustries/forestry/nzstats2000/title.htm

MAF (2002a) New Zealand Forestry information, published on the Ministry of Agriculture and Forestry website, www.maf.govt.nz/forestry/, accessed 25 November 2002

MAF (2002b) *Guidelines for Applicants to the East Coast Forestry Project as at March 2002*, Ministry of Agriculture and Forestry, New Zealand

Malik, R. (1994) *Study of Some Minor Forest Produce in Orissa:An Empirical Investigation*, Nabakrushna Choudhury Centre for Development Studies, Bhubaneshwar.

Marghescu, T. (ed) (2001) Nature Conservation in Private Forests of Selected CEE Countries: Opportunities and Constraints, IUCN, Tilburg, The Netherlands

Mayers, J. and Bass, S. (1999) *Policy that Works for Forests and People*, Policy that Works Series No 7: Series Overview, International Institute for Environment and Development, London

Mayers, J., Bass, S. and Macqueen, D. (2002) *The Pyramid: A Diagnostic and Planning Tool for Good Forest Governance*, Prepared by IIED for the World Bank–WWF Alliance for Forest Conservation and Sustainable Use, Washington, DC

Mayers, J., Evans, J. and Foy, T. (2001) *Raising the Stakes,* International Institute for Environment and Development, London

Mayers, J. and Vermeulen, S. (2002a) *Company–Community Forest Partnerships: From Raw Deals to Mutual Gains?* Instruments for Sustainable Private Sector Forestry Series, International Institute for Environment and Development, London

Mayers, J. and Vermeulen, S. (2002b) *Power from the Trees: How Good Forest Governance Can Help Reduce Poverty,* WSSD Opinion Series, International Institute for Environment and Development, London

MCFFA (Ministerial Council on Forestry, Fisheries and Agriculture) (1997) *Plantations for Australia: the 2020 Vision,* www.plantations2020.com.au/, accessed 31 July 2002

McKenzie Smith, G. R. (1975) *The Private Sector of the Afforestation Industry in Australia with Particular Reference to Afforestation Investment Companies,* MSc thesis (Forest Management), Department of Forestry, Australian National University, April, Australia

MDBC (Murray-Darling Basin Commission) (2002) *A Proposed Vegetation Bank,* Draft Factsheet, MDBC, Canberra, www.mdbc.gov.au/publications/factsheets/fsal004_101.html, accessed 4 August 2002

Mead, D. J. (1995) 'The role of agroforestry in industrialized nations: The southern hemisphere perspective with special emphasis on Australia and New Zealand', *Agroforestry Systems,* vol 31, pp143–156

Miller, R. (ed) (2002) *Impact of Incentives on the Development of Forest Plantation Resources in the Asia-Pacific Region: Australian Case Study,* Report to the 19th Session, Asia-Pacific Forestry Commission, Mongolia, 26–30 August 2002, AFFA, Canberra

Ministry of Forestry (1993) *Forestry Sector Issues: A Post Election Briefing for the Minister of Forestry,* New Zealand Ministry of Forestry, Wellington

Mishra, N. K. (1997) *Social Forestry in India: A Study of Economic Aspects,* Swaraj Prakashan, New Delhi

Morales, E. (2002) 'Changing Ownership and Management of State Forest Plantations. Chile's Experience'. Draft study submitted to the International Conference, Changing Ownership and Management of State Forest Plantations: Issues, Approaches, Implications. Cape Town, South Africa, 6–8 November 2002

Mukherjee, S. D. (2001) 'Status of JFM in Andhra Pradesh'. Draft paper prepared for the book on joint forest management being prepared by the Commonwealth Forestry Association in Collaboration with Winrock International India and Ford Foundation, New Delhi

Murtough, G., Aretino, B. and Matysek, A. (2002) *Creating Markets for Ecosystem Services,* Staff Research Paper, Productivity Commission, Melbourne, www.pc.gov.au/research/staffres/cmfes/index.html, accessed 16 August 2002

Nagashima, K., Sands, R., Whyte, A. D., Bilek, E. M. and Nakagoshi, N. (2002) 'Regional landscape change as a consequence of plantation forestry expansion: An example in the Nelson region, New Zealand', *Forest Ecology and Management,* vol 163(1–3), pp245–261

Nambiar, S. and Brown, A. (eds) (2001) *Plantations, Farm Forestry and Water,* Water and Salinity Issues No 7, RIRDC, Canberra

NCC (National Competition Council) (2002) *Overview of National Competition Policy,* www.ncc.gov.au, accessed 31 July 2002

Neilson, D. (2003) 'Central North Island forests under a jinx', *Rotorua Daily Post,* March 2003

NFI (National Forest Inventory) (1998) *Australia's State of the Forests Report 1998,* Bureau of Rural Sciences, Canberra

NFI (2002) *National Plantation Inventory Tabular Report – March 2002,* Bureau of Rural Sciences, Canberra

NFI (2004) *National Plantation Inventory Update – March 2004,* Bureau of Rural Sciences, Canberra

NRE (2002) 'Forest Rights and Carbon Sequestation Rights', www.nre.vic.gov.au. Accessed 27 July 2002

NRRPC (Northern Rivers Regional Plantation Committee) (2001) 'Northern Rivers Regional Plantation Committee Private Forestry about the NRRPC', www. privateforestry.org.au/n1.htm, accessed 30 July 2002

NZFOA (New Zealand Forest Owners Association) (2002) *Facts and Figures 2002/2003,* www.nzfoa.nzforestry.co.nz, accessed 24 January 2003

NZIER (New Zealand Institute of Economic Research, Inc.) (2000) *Devolving Forest Ownership through Privatisation: Processes, Issues and Outcomes,* Working Paper 2000/3, May 2000, New Zealand Institute of Economic Research, Inc., Wellington

New Zealand Forestry (2004) www.nzforestry.co.nz/ngt-news.asp?articleid=1930. Accessed 14 November, 2004

OECD (Organisation for Economic Co-operation and Development) (1999) 'Government reform: Of roles and functions of government and public corporations', New Zealand Country Paper, Presented at Government of the Future Getting from Here to There ... A Symposium Organized by the OECD, Paris, 14–15 September 1999, www1.oecd.org/puma/strat/symposium/NewZealand.pdf

Office of the Parliamentary Commissioner for the Environment (1994) *Sustainable Land Management and the East Coast Forestry Project,* December 1994, Office of the Parliamentary Commissioner for the Environment, Wellington

Onibon, O., Dabiré, B. and Ferroukhi, L. (2000) 'Local practices and decentralisation and devolution of natural resource management in West Africa: Stakes, challenges and prospects', in Enters, T., Durst, P. B. and Victor, M. (eds) *Decentralisation and Devolution of Forest Management in Asia and the Pacific,* RECOFTC Report No 18 and RAP Publication 2000/2001, Bangkok, Thailand

Pant, M. M. (1979) 'Social forestry in India', *Unasylva,* vol 31, no 125

Palit, S. (2001) 'Overview of JFM status in West Bengal'. Draft paper prepared for a book on joint forest management being planned by Commonwealth Forestry Association (India Chapter) in collaboration with Winrock International India and Ford Foundation, New Delhi

Paperlinx (2001) 'Paperlinx sells Australian paper plantations', Press release, 31 July 2001, www.paperlinx.com.au/main.htm?page=investors_archivedpressreleases, accessed 30 July 2002

Pathak, A (1994) *Contested Domains: The State, Peasants and Forests in Contemporary India,* Sage Publications, New Delhi

Pawson, E. and Scott, G. (1992) 'The regional consequences of economic restructuring: The West Coast, New Zealand (1984–1991)', *Journal of Rural Studies,* vol 8(4), pp373–386

Peacock, S. (2001a) 'Bank puts receivers into APT', *The West Australian,* 1 August 2001, p51

Peacock, S. (2001b) 'Tax break on trees', *The Age,* 10 October 2001, Business Section, www.theage.com.au/business/2001/10/10/FFXUN09LKSC.html, accessed 9 September 2002

Petheram, J., Patterson, A., Williams, K., Jenkin, B. and Nettle, R. (2000) *Socioeconomic Impact of Changing Land Use in South West Victoria,* Institute of Land and Food Resources, University of Melbourne, Melbourne

Phoung, P. X. (2000) 'People's participation in forest management in Vietnam', in Enters, T., Durst, P. B. and Victor, M. (eds) *Decentralisation and Devolution of Forest Management in Asia and the Pacific,* RECOFTC Report No 18 and RAP Publication 2000/2001, Bangkok, Thailand

Planning Commission (1998) *Leasing of Degraded Forest Lands: Working Group's Report on the Prospects of Making Degraded Forests Available to Private Entrepreneurs,* Planning Commission, GoI, New Delhi

Poffenberger, M. and McGean, B. (eds) (1996) *Village Voices, Forest Choices: Joint Forest Management in India,* Oxford University Press, New Delhi

Poffenberger, M. and Sarin. M. (1995) 'Fiber grass from forest land', *Society and Natural Resources,* vol 8

Poole, A. L. (1969) *Forestry in New Zealand: The Shaping of Policy,* Hodder and Stoughton, Auckland

Principles for Commercial Plantation Forest Management in New Zealand (1995) Agreement signed 6 December 1995 between the New Zealand Forest Owners Association Inc., New Zealand Farm Forestry Association, Royal Forest and Bird Protection Society of New Zealand Inc., WWF-New Zealand, Federated Mountain Clubs of New Zealand Inc., and Maruia Society Inc., Nelson

PTAA (2001) 'Industry profile: Australian plantation estate', www.ptaa.com.au/industry/indprof.htm, accessed 30 July 2002

Race, D. (2002) *Innovative Use of Farm Trees: Australian Marketing Experiences,* RIRDC Publication 02/22, Rural Industries Research and Development Corporation, Canberra

Radebe, J. and Kasrils, R. (2000) Ministers for Public Enterprises and Water Services and Forestry, respectively, 'Restructuring of State Forest Assets', 14 September 2000, Cape Town

Rao, G. B., Goswami, A. and Agarwal, C. (1992) *Trends in Social Forestry in India,* Report prepared for the Swedish International Development Authority (SIDA), Society for Promotion of Wastelands Development (SPWD), New Delhi

Roche, M. M. (1987) *Forest Policy in New Zealand: An Historical Geography, 1840–1919,* The Dunmore Press, Palmerston North, New Zealand

Roche, M. (1990a) *History of New Zealand Forestry,* GP Books and New Zealand Forestry Corporation, New Zealand

Roche, M. M. (1990b) 'Perspectives on the post-1984 restructuring of state forestry in New Zealand', *Environment and Planning A,* vol 22, pp941–959

Rockell, J. D. (1974) *Problems of Afforestation,* New Zealand Forest Service, Wellington

Rodger, G. J. (1952) *Report of the Royal Commission Appointed to Inquire into and Report upon Forestry and Timber Matters in Western Australia,* Government Printer, Perth

Roe, D., Mayers, J., Grieg-Gran, M., Kothari, A., Fabricius, C. and Hughes, R. (2000) *Evaluating Eden: Exploring the Myths and Realities of Community-based Wildlife Management,* Series overview, International Institute for Environment and Development, London

Rosoman, G. (1994) *The Plantation Effect: An Ecoforestry Review of the Environmental Effects of Exotic Monoculture Tree Plantations in Aotearoa/New Zealand,* August, Greenpeace New Zealand, Auckland

Routley, R. and Routley, V. (1975) *The Fight for the Forests: The Takeover of Australian Forests for Pines, Wood Chips and Intensive Forestry,* Research School of Social Sciences, Australian National University, Canberra

Royal Forest and Bird Protection Society (2001) *West Coast Forests' Announcement – A Momentous Decision,* Press release, 30 May 2001, www.newsroom.co.nz/story/48890-37-0.html, accessed 9 January 2003

Rozelle, S., Huang, J. and Benzinger, V. (2002) 'Forest exploitation and protection in reform China: Assessing the impacts of policy and economic growth', in Hyde, W., Belcher, B. and Xu, J. (eds) *China's Reforms and Its Forestry Sector: Lessons for Global Forest Policy,* Centre for International Forestry Research, Bogor, Indonesia, Unpublished manuscript

Ruiz Perez, M., Belcher, B., Fu, M. and Yang, X. (2002) 'Forestry, poverty and rural development: Perspectives from the bamboo sub-sector', in Hyde, W., Belcher, B. and Xu, J. (eds) *China's Reforms and Its Forestry Sector: Lessons for Global Forest Policy,* Centre for International Forestry Research, Bogor, Indonesia, Unpublished manuscript

Rule, A. (1967) *Forests of Australia,* Angus and Robertson, Sydney

Saigal, S. (1998) *Participatory Forestry in India: Analysis and Lessons,* MSc thesis, Oxford Forestry Institute, University of Oxford, Oxford

Saigal, S. (2002) 'Changing Ownership and Management of State Forest Plantations: India's Experience'. Draft study submitted to the International Conference, Changing Ownership and Management of State Forest Plantations: Issues, Approaches, Implications. Cape Town, South Africa, 6–8 November 2002

Saigal, S., Arora, H. and Rizvi, S. S. (2002) *The New Foresters: Role of Private Enterprise in the Indian Forestry Sector,* International Institute for Environment and Development, London

Salmon, G. (1993) 'Conservation and environmental management of forests: An NGO's view', *Commonwealth Forestry Review,* vol 72(4), pp233–241

Salvin, S. (2001) 'Carbon, environmental services and planted forests in NSW', *Outlook 2001,* Canberra, February 2001

Sargent, S. (1992) 'Natural forest or plantation', in Sargent, S. and Bass, S. (eds) (1992) *Plantation Politics,* International Institute for Environment and Development, London

Sargent, S. and Bass, S. (eds) (1992) *Plantation Politics,* International Institute for Environment and Development, London

Sayer, J., and Sun, C. (2002) 'The impacts of policy reforms on forest environments and biodiversity', in Hyde, W., Belcher, B. and Xu, J. (eds) *China's Reforms and Its Forestry Sector: Lessons for Global Forest Policy,* Centre for International Forestry Research, Bogor, Indonesia, Unpublished manuscript.

Schirmer, J. (2002a) *Afforestation and Conflict: Social Reactions to Commercial Plantations in South-West of Western Australia,* draft report, The Australian National University, Canberra

Schirmer, J. (2002b) *Plantation Forestry Disputes: Case Studies on Concerns, Causes, Processes and Paths Toward Resolution,* Technical Report No 42 (Revised), Cooperative Research Centre for Sustainable Production Forestry, Hobart

Schirmer, J. and Roche, M. (2003) Changing Ownership and Management of State Forest Plantations: New Zealand/Aoteoroa's Experience'. Unpublished draft study prepared for this book

Schirmer, J. and Kanowski, P. (2002) 'Changing Ownership and Management of State Forest Plantations: Australia's Experience'. Draft Study submitted to the International Conference, Changing Ownership and Management of State Forest Plantations: Issues, Approaches, Implications. Cape Town, South Africa, 6–8 November 2002

Schmitt, G. J. (1972) *Sale of Interests in State Forests: Report to the Honourable the Minister of Forests June 1972*, School of Management Studies, University of Waikato, Hamilton

Scott, K., Park, J. and Cocklin, C. (2000) 'From "sustainable rural communities" to "social sustainability": Giving voice to diversity in Mangakahia Valley, New Zealand', *Journal of Rural Studies*, vol 16(4), pp443–446

SEPA (State Environmental Protection Agency) (1998) *China's Biodiversity:A Country Study,* China Environmental Sciences Press, Beijing

SEPA (1999) *China Environmental Protection Development Report,* China Environmental Science Publishing House, Beijing

SFA (State Forestry Administration) (1949–1987, 1987–2001) *Forestry Statistics Yearbook,* China Forestry Publishing House, Beijing.

SFA (2000a) 'The first to fifth China national forest resource inventory (1973–1998)', in *National Forestry Statistics (1973–1999)*, China Forestry Publishing House, Beijing

SFA (2000b) *China Forestry Development Report,* China Forestry Publishing House, Beijing

SFA (2002) Press conference release for First China Forest Landscape Resource Exhibition and Tianmu Mountain Forest Festival, June 2002

SFNSW (State Forests of New South Wales) (1999) *SFE and State Forests of NSW to Develop World's First Carbon Trading Market,* SFNSW Press Release, www.forest.nsw.gov.au/carbon/media/releases/300899.asp, accessed 06 August 2002

SFNSW (2002) *Planted Forests for Tokyo Electric Power Company (TEPCO),* www.forest.nsw.gov.au/carbon/overview/tepco/default.asp, accessed 30 July 2002

Shea, S. and Bartle, J. (1988) 'Restoring nature's balance: The potential for major reforestation of South Western Australia', Reprinted from *LANDSCOPE,* vol 3(3) Department of CALM, WA

Simpson, T.E. (1973) *Kauri to Radiata: Origin and Expansion of the Timber Industry of New Zealand,* Hodder and Stoughton, Auckland

Singh, C. (1986) *Common Property and Common Poverty: India's Forests, Forest Dwellers and the Law,* Oxford University Press, Delhi

Singh, H. B. (2000) 'Community forestry implementation: Emerging institutional linkages', in Enters, T., Durst, P. B. and Victor, M. (eds) *Decentralisation and Devolution of Forest management in Asia and the Pacific,* RECOFTC Report No 18 and RAP Publication 2000/2001, Bangkok, Thailand

Singhania, H. S. (1997) *Presentation Made to the Honourable Union Industry Minister, GoI, by Shri Hari Shankar Singhania, Chairman of the Committee to Suggest an Action Plan for Pulp and Paper Industry,* 28 May

Smartwood (2002) 'Public Briefing Note for FSC Forest Management Certification Assessment of Hancock Victorian Plantations Pty Ltd, Australia', September, www.certifiedforests.org.au/homef.html >Documents, accessed 15 August 2002

Snell, A. and Vise, S. (eds) (2000) *Opportunities for the New Millennium,* Proceedings of the Australian Forest Growers Biennial Conference, Cairns, pp191–195

SPIS (State Plantations Impact Study) (1990) *State Plantations Impact Study: Report and Recommendations,* State Plantations Impact Study Steering Committee, Melbourne

Standing Committee on Environment and Conservation (1975) *The Operation of the Softwood Forestry Agreements Acts 1967 and 1972: Report from the House of Representatives Standing Committee on Environment and Conservation,* May 1975, AGPS, Canberra

Stanton, R. J. (2000) *Plantations 2020 Vision – Progress Report*, Report to Standing Committee on Forestry of the Ministerial Council on Agriculture, Fisheries and Forestry, Canberra

Stayner, R. (1999) *Value-adding to Regional Communities and Farming Industries*, Background paper, Regional Australia Summit, 27–29 October, www.dotrs.gov.au/regional/summit/background/index.htm, accessed 30 July 2002

Stirzaker,. R, Vertessy, R. and Sarre, A. (eds) (2002) *Trees, Water and Salt*, RIRDC Publication 01/086, RIRDC, Canberra

Storey, T. and Clayton, B. (2002) 'Environmental law regime in New Zealand', *New Zealand Journal of Forestry*, vol 46(4), pp41–43

Storey, T. and Taylor, C. (2001) 'Regulating overseas investment in forestry', *New Zealand Journal of Forestry*, vol 46(1), pp38–40

Stringer, C. A. (2002) 'Rescaling the region: The integration of Hawkes Bay food and forestry industries into East Asia', *Australian Geographer*, vol 33(1), pp63–77

Swale, B. (2001) 'New Zealand government sales of state assets continue despite policy to contrary. Covert government actions call into question the integrity of the present Labour government', Press Release issued Monday, 16 July 2001, www.homepages.caverock.net.nz/~bj/beech/pressrel/press29.htm, accessed 25 November 2002

Tewari, D. N. (1995) *Marketing and Trade of Forest Produce*, International Book Distributors, Dehradun

Tomlinson, C. J., Fairweather, J. R. and Swaffield, S. R. (2000) *Gisborne/East Coast Field Research on Attitudes to Land Use Change: An Analysis of Impediments to Forest Sector Development*, Research Report No 249, December 2000, Agribusiness and Economics Research Unit, Lincoln University, Canterbury

Tonts, M., Campbell, C. and Black, A. (2001) *Socio-Economic Impacts of Farm Forestry*, Report for the RIRDC/LWRRDC/FWPRDC Joint Venture Agroforestry Programme, May 2001, Rural Industries Research and Development Corporation, Canberra

Trembath, A. (2002). *Competitive Neutrality: Scope for Enhancement*, National Competition Council Staff Discussion Paper, AusInfo, Canberra

Tuckey, W. (2000) 'Tuckey announces $450,000 in funding for Regional Plantation Committees in WA', AFFA Media Release 5 October 2000, www.affa.gov.au/ministers/tuckey/releases/00/00_70tu.html, accessed 30 July 2002

Turner, J. A., Buongiorno, J., Horgan, G. P. and Maplesden, F. M. (2001) 'Liberalisation of forestry product trade and the New Zealand forest sector, 2000–2015: A global modelling approach', *New Zealand Journal of Forestry Science*, vol 31(3), pp320–338

UN (United Nations) (1992a) *Non-Legally Binding Authoritative Statement of Principles for a Global Consensus on the Management, Conservation and Sustainable Development on All Types of Forests*, www.un.org/documents/ga/conf151/aconf15126-3annex3.htm

UN (1992b) *Agenda 21*, Chapter 11, 'Combating deforestation', www.un.org/esa/sustdev/agenda21chapter11.htm

Upreti, B. P. and Shrestha, B. P. (2000) 'Balancing power in community forestry: decentralisation and devolution of power', in Enters, T., Durst, P. B. and Victor, M. (eds) (2000) *Decentralisation and Devolution of Forest Management in Asia and the Pacific*, RECOFTC Report No 18 and RAP Publication 2000/2001, Bangkok, Thailand

van Bueren, M. (2001) *Emerging Markets for Environmental Services: Implications and Opportunities for Resource Management in Australia*, Report for the RIRDC/Land

and Water Australia/FWPRDC Joint Venture Agroforestry Programme, RIRDC Publication No 01/162, RIRDC, Canberra

Victorian Auditor-General's Office (Australia) (1999) *Report on Ministerial Portfolios May 1999*, VAGO, Victoria, www.audit.vic.gov.au/old/mp99/mp99cv.htm.

Vira, B. (1995) *Institutional Change in India's Forest Sector, 1976–1994: Reflections on State Policy*, Oxford Centre for the Environment, Ethics and Society (OCEES) Research Paper No 5, OCEES, Oxford

Visser, R. (1996) 'Improving environmental performance: the role of the New Zealand Forest Code of Practice', *New Zealand Journal of Forestry Science*, vol 26(2), pp158–162

Waggener, T. (1998) *Status of Forest Products Pricing under Reforms Towards Market Economies: China, Mongolia, Myanmar and Vietnam*, Final report for FAO/UN support to the reorientation of forestry policies and institutions of countries of Asia in reform to the market economy, FAO/UN, Rome

Walker, B. and Walker, B. C. (2000) *Privatisation Sell Off or Sell Out? The Australian Experience*, ABC Books, Sydney

Walker, L., Cocklin, C. and Le Heron, R. (2000) 'Regulating for environmental improvement in the New Zealand forestry sector', *Geoforum*, vol 31, pp281–297

Wareing, K., Pigson, B., Poynter, M., Carruthers, G. and Baker, R. (2002) *The Timber Industry in North East Victoria; A Socio-economic Assessment*, Report for Plantations North East Inc. Prospect Consulting

West Coast Environmental Law (1999) 'Save BC's public lands: Backgrounder on privatisation–environment issues', www.wcel.org/wcelpub/1999

Wijewardana, D. (2000) 'The impact of policies and legislation on plantation forestry development in New Zealand', Paper to the International Conference on Timber Plantation Development, 7–9 November 2000, Manila, the Philippines

Wilkinson, G. R. (1999) 'Codes of forest practice as regulatory tools for sustainable forest management', Paper presented to the 18th Biennial Conference of the Institute of Foresters of Australia, Hobart, Tasmania, 3–8 October 1999

Williams, J et al (2001) *The Contribution of Mid- to Low-rainfall Forestry and Agroforestry to Contribute to Greenhouse and Natural Resource Management Outcomes*, AGO/ MDBC, www.greenhouse.gov.au/land/gh_land/pubs/abs_lowrainfall.html

Wilton, S. A. R. (2002) *Taxation Treatment of Non-resident Partnership Investors*, 18 October 2001, www.forestenterprises.co.nz/exist/oii/tax_nonresidents_partnership. pdf, accessed 6 April 2003

Wood, M. S., Stephens, N. C., Allison, B. K. and Howell, C. I. (2001) *Plantations of Australia – A Report from the National Plantation Inventory and the National Farm Forest Inventory*, National Forest Inventory, Bureau of Rural Sciences, Canberra

World Bank (1993) *India: Policies and Issues*, Forest Sector Development Report, No 10965-IN, World Bank, Washington, DC

WRI (World Resources Institute), UNEP (United Nations Environment Programme) and UNDP (United Nations Development Programme) (1994) *World Resources 1994–1995: A Guide to the Global Environment*, Oxford University Press, New York

Xu, J. (1999) *China's Paper Industry: Growth and Environmental Policy During Economic Reform*, PhD thesis, Virginia Polytechnic Institute and State University, Virginia.

Xu, J. and Hyde, W. (2002) 'Changing Ownership and Management of State Forest Plantations: China's Experience'. Draft Study submitted to the International Conference, Changing Ownership and Management of State Forest Plantations: Issues, Approaches, Implications. Cape Town, South Africa, 6–8 November 2002

Yin, R. S. and Hyde, W. F. (2000) 'The impact of agroforestry on agricultural productivity: The case of Northern China', *Agroforestry Systems*, vol 50, pp179–194

Yin, R. and Newman, D. H. (1997) 'Impacts of rural reforms: The case of the Chinese forest sector', *Environment and Development Economics*, vol 2, pp291–305

Zhang, D. (2002) 'Policy reform and investment in forestry', in Hyde, W., Belcher, B. and Xu, J. (eds) *China's Reforms and Its Forestry Sector: Lessons for Global Forest Policy*, Centre for International Forestry Research, Bogor, Indonesia, Unpublished manuscript

Zhang, Y (2001) Economic Transition and Forestry Development, PhD thesis, Department of Forest Economics, University of Helsinki, Helsinki

Zhang, Y., Dai, G., Huang, H., Fanwen, K., Tian, Z., Wang, X. and Zhang, L. (1994) 'The forest sector in China', in Palu, M. and Uusivuori, J. (eds) *World Forests, Society, and Environment*, vol 1, Kluwer Academic Publishers, Dordrecht

Zhang, Y., Uusivuori, J. and Kuuluvainen, J. (2000) 'Econometric analysis of the causes of forestland use/cover change in Hainan, China', *Canadian Journal of Forest Research*, vol 30, pp1913–1921

Index